Java核心技术
卷I：基础知识 下
（第 11 版 · 英文版）

美] 凯·S. 霍斯特曼（Cay S. Horstmann） 著

Core Java
Volume I—Fundamentals, Eleventh Edition

人民邮电出版社

北京

目录

CHAPTER 8

Generic Programming

In this chapter

Generic classes and methods have type parameters. This allows them to describe precisely what should happen when they are instantiated with specific types. Prior to generic classes, programmers had to use the Object for writing code that works with multiple types. This was both cumbersome and unsafe.

With the introduction of generics, Java has an expressive type system that allows designers to describe in detail how types of variables and methods should vary. In straightforward situations, you will find it simple to implement generic code. In more advanced cases, it can get quite complex—for implemetors. The goal is to provide classes and methods that other programmers can use without surprises.

The introduction of generics in Java 5 constitutes the most significant change in the Java programming language since its initial release. A major design goal was to be backwards compatible with earlier releases. As a result, Java generics have some uncomfortable limitations. You will learn about the benefits and challenges of generic programming in this chapter.

8.1 Why Generic Programming?

Generic programming means writing code that can be reused for objects of many different types. For example, you don't want to program separate classes to collect String and File objects. And you don't have to—the single class ArrayList collects objects of any class. This is one example of generic programming.

Actually, Java had an ArrayList class before it had generic classes. Let us investigate how the mechanism for generic programming has evolved, and what that means for users and implementors.

8.1.1 The Advantage of Type Parameters

Before generic classes were added to Java, generic programming was achieved with *inheritance*. The ArrayList class simply maintained an array of Object references:

```
public class ArrayList // before generic classes
{
    private Object[] elementData;
    . . .
    public Object get(int i) { . . . }
    public void add(Object o) { . . . }
}
```

This approach has two problems. A cast is necessary whenever you retrieve a value:

```
ArrayList files = new ArrayList();
. . .
String filename = (String) files.get(0);
```

Moreover, there is no error checking. You can add values of any class:

```
files.add(new File(". . ."));
```

This call compiles and runs without error. Elsewhere, casting the result of get to a String will cause an error.

Generics offer a better solution: *type parameters*. The ArrayList class now has a type parameter that indicates the element type:

```
var files = new ArrayList<String>();
```

This makes your code easier to read. You can tell right away that this particular array list contains String objects.

 NOTE: If you declare a variable with an explicit type instead of var, you can omit the type parameter in the constructor by using the "diamond" syntax:

```
ArrayList<String> files = new ArrayList<>();
```

The omitted type is inferred from the type of the variable.

Java 9 expands the use of the diamond syntax to situations where it was previously not accepted. For example, you can now use diamonds with anonymous subclasses:

```
ArrayList<String> passwords = new ArrayList<>() // diamond OK in Java 9
    {
        public String get(int n) { return super.get(n).replaceAll(".", "*"); }
    };
```

The compiler can make good use of the type information too. No cast is required for calling get. The compiler knows that the return type is String, not Object:

```
String filename = files.get(0);
```

The compiler also knows that the add method of an ArrayList<String> has a parameter of type String. That is a lot safer than having an Object parameter. Now the compiler can check that you don't insert objects of the wrong type. For example, the statement

```
files.add(new File(". . .")); // can only add String objects to an ArrayList<String>
```

will not compile. A compiler error is much better than a class cast exception at runtime.

This is the appeal of type parameters: They make your programs easier to read and safer.

8.1.2 Who Wants to Be a Generic Programmer?

It is easy to use a generic class such as ArrayList. Most Java programmers will simply use types such as ArrayList<String> as if they had been built into the language, just like String[] arrays. (Of course, array lists are better than arrays because they can expand automatically.)

However, it is not so easy to implement a generic class. The programmers who use your code will want to plug in all sorts of classes for your type parameters. They will expect everything to work without onerous restrictions and confusing error messages. Your job as a generic programmer, therefore, is to anticipate all the potential future uses of your class.

How hard can this get? Here is a typical issue that the designers of the standard class library had to grapple with. The ArrayList class has a method addAll to add all elements of another collection. A programmer may want to add all elements from an ArrayList<Manager> to an ArrayList<Employee>. But, of course, doing it the other way round should not be legal. How do you allow one call and disallow the other? The Java language designers invented an ingenious new concept, the *wildcard type*, to solve this problem. Wildcard types are rather abstract, but they allow a library builder to make methods as flexible as possible.

Generic programming falls into three skill levels. At a basic level, you just use generic classes—typically, collections such as ArrayList—without thinking how and why they work. Most application programmers will want to stay at that level until something goes wrong. You may, however, encounter a confusing error message when mixing different generic classes, or when interfacing with legacy code that knows nothing about type parameters; at that point, you'll need to learn enough about Java generics to solve problems systematically rather than through random tinkering. Finally, of course, you may want to implement your own generic classes and methods.

Application programmers probably won't write lots of generic code. The JDK developers have already done the heavy lifting and supplied type parameters for all the collection classes. As a rule of thumb, only code that traditionally involved lots of casts from very general types (such as Object or the Comparable interface) will benefit from using type parameters.

In this chapter, we will show you everything you need to know to implement your own generic code. However, we expect that most readers will use this knowledge primarily for help with troubleshooting and to satisfy their curiosity about the inner workings of the parameterized collection classes.

8.2 Defining a Simple Generic Class

A *generic class* is a class with one or more type variables. In this chapter, we will use a simple Pair class as an example. This class allows us to focus on generics without being distracted by data storage details. Here is the code for the generic Pair class:

```
public class Pair<T>
{
   private T first;
   private T second;

   public Pair() { first = null; second = null; }
   public Pair(T first, T second) { this.first = first; this.second = second; }

   public T getFirst() { return first; }
   public T getSecond() { return second; }

   public void setFirst(T newValue) { first = newValue; }
   public void setSecond(T newValue) { second = newValue; }
}
```

The Pair class introduces a type variable T, enclosed in angle brackets < >, after the class name. A generic class can have more than one type variable. For example, we could have defined the Pair class with separate types for the first and second field:

```
public class Pair<T, U> { . . . }
```

The type variables are used throughout the class definition to specify method return types and the types of fields and local variables. For example:

```
private T first; // uses the type variable
```

 NOTE: It is common practice to use uppercase letters for type variables, and to keep them short. The Java library uses the variable E for the element type of a collection, K and V for key and value types of a table, and T (and the neighboring letters U and S, if necessary) for "any type at all."

You *instantiate* the generic type by substituting types for the type variables, such as

```
Pair<String>
```

You can think of the result as an ordinary class with constructors

```
Pair<String>()
Pair<String>(String, String)
```

and methods

```
String getFirst()
String getSecond()
```

```
void setFirst(String)
void setSecond(String)
```

In other words, the generic class acts as a factory for ordinary classes.

The program in Listing 8.1 puts the Pair class to work. The static minmax method traverses an array and simultaneously computes the minimum and maximum values. It uses a Pair object to return both results. Recall that the compareTo method compares two strings, returning 0 if the strings are identical, a negative integer if the first string comes before the second in dictionary order, and a positive integer otherwise.

 C++ NOTE: Superficially, generic classes in Java are similar to template classes in C++. The only obvious difference is that Java has no special template keyword. However, as you will see throughout this chapter, there are substantial differences between these two mechanisms.

Listing 8.1 pair1/PairTest1.java

```
1  package pair1;
2
3  /**
4   * @version 1.01 2012-01-26
5   * @author Cay Horstmann
6   */
7  public class PairTest1
8  {
9     public static void main(String[] args)
10    {
11       String[] words = { "Mary", "had", "a", "little", "lamb" };
12       Pair<String> mm = ArrayAlg.minmax(words);
13       System.out.println("min = " + mm.getFirst());
14       System.out.println("max = " + mm.getSecond());
15    }
16 }
17
18 class ArrayAlg
19 {
20    /**
21     * Gets the minimum and maximum of an array of strings.
22     * @param a an array of strings
23     * @return a pair with the min and max values, or null if a is null or empty
24     */
25    public static Pair<String> minmax(String[] a)
26    {
27       if (a == null || a.length == 0) return null;
28       String min = a[0];
```

```
29        String max = a[0];
30        for (int i = 1; i < a.length; i++)
31        {
32           if (min.compareTo(a[i]) > 0) min = a[i];
33           if (max.compareTo(a[i]) < 0) max = a[i];
34        }
35        return new Pair<>(min, max);
36     }
37  }
```

8.3 Generic Methods

In the preceding section, you have seen how to define a generic class. You can also define a single method with type parameters.

```
class ArrayAlg
{
   public static <T> T getMiddle(T... a)
   {
      return a[a.length / 2];
   }
}
```

This method is defined inside an ordinary class, not inside a generic class. However, it is a generic method, as you can see from the angle brackets and the type variable. Note that the type variables are inserted after the modifiers (public static, in our case) and before the return type.

You can define generic methods both inside ordinary classes and inside generic classes.

When you call a generic method, you can place the actual types, enclosed in angle brackets, before the method name:

```
String middle = ArrayAlg.<String>getMiddle("John", "Q.", "Public");
```

In this case (and indeed in most cases), you can omit the <String> type parameter from the method call. The compiler has enough information to infer the method that you want. It matches the type of the arguments against the generic type T... and deduces that T must be String. That is, you can simply call

```
String middle = ArrayAlg.getMiddle("John", "Q.", "Public");
```

In almost all cases, type inference for generic methods works smoothly. Occasionally, the compiler gets it wrong, and you'll need to decipher an error report. Consider this example:

```
double middle = ArrayAlg.getMiddle(3.14, 1729, 0);
```

The error message complains, in cryptic terms that vary from one compiler version to another, that there are two ways of interpreting this code, both equally valid. In a nutshell, the compiler autoboxed the parameters into a Double and two Integer objects, and then it tried to find a common supertype of these classes. It actually found two: Number and the Comparable interface, which is itself a generic type. In this case, the remedy is to write all parameters as double values.

TIP: Peter von der Ahé recommends this trick if you want to see which type the compiler infers for a generic method call: Purposefully introduce an error and study the resulting error message. For example, consider the call ArrayAlg.getMiddle("Hello", 0, null). Assign the result to a JButton, which can't possibly be right. You will get an error report:

```
found:
java.lang.Object&java.io.Serializable&java.lang.Comparable<? extends
java.lang.Object&java.io.Serializable&java.lang.Comparable<?>>
```

In plain English, you can assign the result to Object, Serializable, or Comparable.

C++ NOTE: In C++, you place the type parameters after the method name. That can lead to nasty parsing ambiguities. For example, g(f<a,b>(c)) can mean "call g with the result of f<a,b>(c)", or "call g with the two boolean values f<a and b>(c)".

8.4 Bounds for Type Variables

Sometimes, a class or a method needs to place restrictions on type variables. Here is a typical example. We want to compute the smallest element of an array:

```
class ArrayAlg
{
   public static <T> T min(T[] a) // almost correct
   {
      if (a == null || a.length == 0) return null;
      T smallest = a[0];
      for (int i = 1; i < a.length; i++)
         if (smallest.compareTo(a[i]) > 0) smallest = a[i];
      return smallest;
   }
}
```

But there is a problem. Look inside the code of the min method. The variable smallest has type T, which means it could be an object of an arbitrary class. How do we know that the class to which T belongs has a compareTo method?

The solution is to restrict T to a class that implements the Comparable interface—a standard interface with a single method, compareTo. You can achieve this by giving a *bound* for the type variable T:

```
public static <T extends Comparable> T min(T[] a) . . .
```

Actually, the Comparable interface is itself a generic type. For now, we will ignore that complexity and the warnings that the compiler generates. Section 8.8, "Wildcard Types," on p. 459 discusses how to properly use type parameters with the Comparable interface.

Now, the generic min method can only be called with arrays of classes that implement the Comparable interface, such as String, LocalDate, and so on. Calling min with a Rectangle array is a compile-time error because the Rectangle class does not implement Comparable.

 C++ NOTE: In C++, you cannot restrict the types of template parameters. If a programmer instantiates a template with an inappropriate type, an (often obscure) error message is reported inside the template code.

You may wonder why we use the extends keyword rather than the implements keyword in this situation—after all, Comparable is an interface. The notation

```
<T extends BoundingType>
```

expresses that T should be a *subtype* of the bounding type. Both T and the bounding type can be either a class or an interface. The extends keyword was chosen because it is a reasonable approximation of the subtype concept, and the Java designers did not want to add a new keyword (such as sub) to the language.

A type variable or wildcard can have multiple bounds. For example:

```
T extends Comparable & Serializable
```

The bounding types are separated by ampersands (&) because commas are used to separate type variables.

As with Java inheritance, you can have as many interface supertypes as you like, but at most one of the bounds can be a class. If you have a class as a bound, it must be the first one in the bounds list.

In the next sample program (Listing 8.2), we rewrite the minmax method to be generic. The method computes the minimum and maximum of a generic array, returning a Pair<T>.

Listing 8.2 pair2/PairTest2.java

```
1  package pair2;
2
3  import java.time.*;
4
5  /**
6   * @version 1.02 2015-06-21
7   * @author Cay Horstmann
8   */
9  public class PairTest2
10 {
11    public static void main(String[] args)
12    {
13       LocalDate[] birthdays =
14          {
15             LocalDate.of(1906, 12, 9), // G. Hopper
16             LocalDate.of(1815, 12, 10), // A. Lovelace
17             LocalDate.of(1903, 12, 3), // J. von Neumann
18             LocalDate.of(1910, 6, 22), // K. Zuse
19          };
20       Pair<LocalDate> mm = ArrayAlg.minmax(birthdays);
21       System.out.println("min = " + mm.getFirst());
22       System.out.println("max = " + mm.getSecond());
23    }
24 }
25
26 class ArrayAlg
27 {
28    /**
29       Gets the minimum and maximum of an array of objects of type T.
30       @param a an array of objects of type T
31       @return a pair with the min and max values, or null if a is null or empty
32    */
33    public static <T extends Comparable> Pair<T> minmax(T[] a)
34    {
35       if (a == null || a.length == 0) return null;
36       T min = a[0];
37       T max = a[0];
38       for (int i = 1; i < a.length; i++)
39       {
40          if (min.compareTo(a[i]) > 0) min = a[i];
41          if (max.compareTo(a[i]) < 0) max = a[i];
42       }
```

```
43        return new Pair<>(min, max);
44    }
45 }
```

8.5 Generic Code and the Virtual Machine

The virtual machine does not have objects of generic types—all objects belong to ordinary classes. An earlier version of the generics implementation was even able to compile a program that used generics into class files that executed on 1.0 virtual machines! In the following sections, you will see how the compiler "erases" type parameters, and what implication that process has for Java programmers.

8.5.1 Type Erasure

Whenever you define a generic type, a corresponding *raw* type is automatically provided. The name of the raw type is simply the name of the generic type, with the type parameters removed. The type variables are *erased* and replaced by their bounding types (or Object for variables without bounds).

For example, the raw type for Pair<T> looks like this:

```
public class Pair
{
   private Object first;
   private Object second;

   public Pair(Object first, Object second)
   {
      this.first = first;
      this.second = second;
   }

   public Object getFirst() { return first; }
   public Object getSecond() { return second; }

   public void setFirst(Object newValue) { first = newValue; }
   public void setSecond(Object newValue) { second = newValue; }
}
```

Since T is an unbounded type variable, it is simply replaced by Object.

The result is an ordinary class, just as you might have implemented it before generics were added to Java.

Your programs may contain different kinds of Pair, such as Pair<String> or Pair<LocalDate>, but erasure turns them all into raw Pair types.

C++ NOTE: In this regard, Java generics are very different from C++ templates. C++ produces different types for each template instantiation—a phenomenon called "template code bloat." Java does not suffer from this problem.

The raw type replaces type variables with the first bound, or Object if no bounds are given. For example, the type variable in the class Pair<T> has no explicit bounds, hence the raw type replaces T with Object. Suppose we declare a slightly different type:

```
public class Interval<T extends Comparable & Serializable> implements Serializable
{
   private T lower;
   private T upper;
   . . .
   public Interval(T first, T second)
   {
      if (first.compareTo(second) <= 0) { lower = first; upper = second; }
      else { lower = second; upper = first; }
   }
}
```

The raw type Interval looks like this:

```
public class Interval implements Serializable
{
   private Comparable lower;
   private Comparable upper;
   . . .
   public Interval(Comparable first, Comparable second) { . . . }
}
```

 NOTE: You may wonder what happens if you switch the bounds: class Interval<T extends Serializable & Comparable>. In that case, the raw type replaces T with Serializable, and the compiler inserts casts to Comparable when necessary. For efficiency, you should therefore put tagging interfaces (that is, interfaces without methods) at the end of the bounds list.

8.5.2 Translating Generic Expressions

When you program a call to a generic method, the compiler inserts casts when the return type has been erased. For example, consider the sequence of statements

```
Pair<Employee> buddies = . . .;
Employee buddy = buddies.getFirst();
```

The erasure of getFirst has return type Object. The compiler automatically inserts the cast to Employee. That is, the compiler translates the method call into two virtual machine instructions:

- A call to the raw method Pair.getFirst
- A cast of the returned Object to the type Employee

Casts are also inserted when you access a generic field. Suppose the first and second fields of the Pair class were public. (Not a good programming style, perhaps, but it is legal Java.) Then the expression

```
Employee buddy = buddies.first;
```

also has a cast inserted in the resulting bytecodes.

8.5.3 Translating Generic Methods

Type erasure also happens for generic methods. Programmers usually think of a generic method such as

```
public static <T extends Comparable> T min(T[] a)
```

as a whole family of methods, but after erasure, only a single method is left:

```
public static Comparable min(Comparable[] a)
```

Note that the type parameter T has been erased, leaving only its bounding type Comparable.

Erasure of methods brings up a couple of complexities. Consider this example:

```
class DateInterval extends Pair<LocalDate>
{
   public void setSecond(LocalDate second)
   {
      if (second.compareTo(getFirst()) >= 0)
         super.setSecond(second);
   }
   . . .
}
```

A date interval is a pair of LocalDate objects, and we'll want to override the methods to ensure that the second value is never smaller than the first. This class is erased to

```
class DateInterval extends Pair // after erasure
{
   public void setSecond(LocalDate second) { . . . }
   . . .
}
```

Perhaps surprisingly, there is another setSecond method, inherited from Pair, namely

```
public void setSecond(Object second)
```

This is clearly a different method because it has a parameter of a different type—Object instead of LocalDate. But it *shouldn't* be different. Consider this sequence of statements:

```
var interval = new DateInterval(. . .);
Pair<LocalDate> pair = interval; // OK--assignment to superclass
pair.setSecond(aDate);
```

Our expectation is that the call to setSecond is polymorphic and that the appropriate method is called. Since pair refers to a DateInterval object, that should be DateInterval.setSecond. The problem is that the type erasure interferes with polymorphism. To fix this problem, the compiler generates a *bridge method* in the DateInterval class:

```
public void setSecond(Object second) { setSecond((LocalDate) second); }
```

To see why this works, let us carefully follow the execution of the statement

```
pair.setSecond(aDate)
```

The variable pair has declared type Pair<LocalDate>, and that type only has a single method called setSecond, namely setSecond(Object). The virtual machine calls that method on the object to which pair refers. That object is of type DateInterval. Therefore, the method DateInterval.setSecond(Object) is called. That method is the synthesized bridge method. It calls DateInterval.setSecond(LocalDate), which is what we want.

Bridge methods can get even stranger. Suppose the DateInterval class also overrides the getSecond method:

```
class DateInterval extends Pair<LocalDate>
{
   public LocalDate getSecond() { return (LocalDate) super.getSecond(); }
   . . .
}
```

In the DateInterval class, there are two getSecond methods:

```
LocalDate getSecond() // defined in DateInterval
Object getSecond() // overrides the method defined in Pair to call the first method
```

You could not write Java code like that; it would be illegal to have two methods with the same parameter types—here, with no parameters. However, in the virtual machine, the parameter types *and the return type* specify a method. Therefore, the compiler can produce bytecodes for two methods that differ only in their return type, and the virtual machine will handle this situation correctly.

 NOTE: Bridge methods are not limited to generic types. We already noted in Chapter 5 that it is legal for a method to specify a more restrictive return type when overriding another method. For example:

```
public class Employee implements Cloneable
{
    public Employee clone() throws CloneNotSupportedException { . . . }
}
```

The Object.clone and Employee.clone methods are said to have *covariant return types*.

Actually, the Employee class has *two* clone methods:

```
Employee clone() // defined above
Object clone() // synthesized bridge method, overrides Object.clone
```

The synthesized bridge method calls the newly defined method.

In summary, you need to remember these facts about translation of Java generics:

- There are no generics in the virtual machine, only ordinary classes and methods.
- All type parameters are replaced by their bounds.
- Bridge methods are synthesized to preserve polymorphism.
- Casts are inserted as necessary to preserve type safety.

8.5.4 Calling Legacy Code

When Java generics were designed, a major goal was to allow interoperability between generics and legacy code. Let us look at a concrete example of such legacy. The Swing user interface toolkit provides a JSlider class whose "ticks" can be customized with labels that contain text or images. The labels are set with the call

```
void setLabelTable(Dictionary table)
```

The Dictionary class maps integers to labels. Before Java 5, that class was implemented as a map of Object instances. Java 5 made Dictionary into a generic class, but JSlider was never updated. At this point, Dictionary without type parameters is a raw type. This is where compatibility comes in.

When you populate the dictionary, you can use the generic type.

```
Dictionary<Integer, Component> labelTable = new Hashtable<>();
labelTable.put(0, new JLabel(new ImageIcon("nine.gif")));
labelTable.put(20, new JLabel(new ImageIcon("ten.gif")));
. . .
```

When you pass the Dictionary<Integer, Component> object to setLabelTable, the compiler issues a warning.

```
slider.setLabelTable(labelTable); // warning
```

After all, the compiler has no assurance about what the setLabelTable might do to the Dictionary object. That method might replace all the keys with strings. That breaks the guarantee that the keys have type Integer, and future operations may cause bad cast exceptions.

You should ponder it and ask what the JSlider is actually going to do with this Dictionary object. In our case, it is pretty clear that the JSlider only reads the information, so we can ignore the warning.

Now consider the opposite case, in which you get an object of a raw type from a legacy class. You can assign it to a variable whose type uses generics, but of course you will get a warning. For example:

```
Dictionary<Integer, Components> labelTable = slider.getLabelTable(); // warning
```

That's OK—review the warning and make sure that the label table really contains Integer and Component objects. Of course, there never is an absolute guarantee. A malicious coder might have installed a different Dictionary in the slider. But again, the situation is no worse than it was before generics. In the worst case, your program will throw an exception.

After you are done pondering the warning, you can use an *annotation* to make it disappear. You can annotate a local variable:

```
@SuppressWarnings("unchecked")
Dictionary<Integer, Components> labelTable = slider.getLabelTable(); // no warning
```

Or you can annotate an entire method, like this:

```
@SuppressWarnings("unchecked")
public void configureSlider() { . . . }
```

This annotation turns off checking for all code inside the method.

8.6 Restrictions and Limitations

In the following sections, we discuss a number of restrictions that you need to consider when working with Java generics. Most of these restrictions are a consequence of type erasure.

8.6.1 Type Parameters Cannot Be Instantiated with Primitive Types

You cannot substitute a primitive type for a type parameter. Thus, there is no Pair<double>, only Pair<Double>. The reason is, of course, type erasure. After erasure, the Pair class has fields of type Object, and you can't use them to store double values.

This is an annoyance, to be sure, but it is consistent with the separate status of primitive types in the Java language. It is not a fatal flaw—there are only eight primitive types, and you can always handle them with separate classes and methods when wrapper types are not an acceptable substitute.

8.6.2 Runtime Type Inquiry Only Works with Raw Types

Objects in the virtual machine always have a specific nongeneric type. Therefore, all type inquiries yield only the raw type. For example,

```
if (a instanceof Pair<String>) // ERROR
```

could only test whether a is a Pair of any type. The same is true for the test

```
if (a instanceof Pair<T>) // ERROR
```

or the cast

```
Pair<String> p = (Pair<String>) a; // warning--can only test that a is a Pair
```

To remind you of the risk, you will get a compiler error (with instanceof) or warning (with casts) when you try to inquire whether an object belongs to a generic type.

In the same spirit, the getClass method always returns the raw type. For example:

```
Pair<String> stringPair = . . .;
Pair<Employee> employeePair = . . .;
if (stringPair.getClass() == employeePair.getClass()) // they are equal
```

The comparison yields true because both calls to getClass return Pair.class.

8.6.3 You Cannot Create Arrays of Parameterized Types

You cannot instantiate arrays of parameterized types, such as

```
var table = new Pair<String>[10]; // ERROR
```

What's wrong with that? After erasure, the type of `table` is `Pair[]`. You can convert it to `Object[]`:

```
Object[] objarray = table;
```

An array remembers its component type and throws an `ArrayStoreException` if you try to store an element of the wrong type:

```
objarray[0] = "Hello"; // ERROR--component type is Pair
```

But erasure renders this mechanism ineffective for generic types. The assignment

```
objarray[0] = new Pair<Employee>();
```

would pass the array store check but still result in a type error. For this reason, arrays of parameterized types are outlawed.

Note that only the creation of these arrays is outlawed. You can declare a variable of type `Pair<String>[]`. But you can't initialize it with a `new Pair<String>[10]`.

NOTE: You can declare arrays of wildcard types and then cast them:

```
var table = (Pair<String>[]) new Pair<?>[10];
```

The result is not safe. If you store a `Pair<Employee>` in `table[0]` and then call a `String` method on `table[0].getFirst()`, you get a `ClassCastException`.

TIP: If you need to collect parameterized type objects, simply use an `ArrayList`: `ArrayList<Pair<String>>` is safe and effective.

8.6.4 Varargs Warnings

In the preceding section, you saw that Java doesn't support arrays of generic types. In this section, we discuss a related issue: passing instances of a generic type to a method with a variable number of arguments.

Consider this simple method with variable arguments:

```
public static <T> void addAll(Collection<T> coll, T... ts)
{
    for (T t : ts) coll.add(t);
}
```

Recall that the parameter ts is actually an array that holds all supplied arguments.

Now consider this call:

```
Collection<Pair<String>> table = . . .;
Pair<String> pair1 = . . .;
Pair<String> pair2 = . . .;
addAll(table, pair1, pair2);
```

In order to call this method, the Java virtual machine must make an array of Pair<String>, which is against the rules. However, the rules have been relaxed for this situation, and you only get a warning, not an error.

You can suppress the warning in one of two ways. You can add the annotation @SuppressWarnings("unchecked") to the method containing the call to addAll. Or, as of Java 7, you can annotate the addAll method itself with @SafeVarargs:

```
@SafeVarargs
public static <T> void addAll(Collection<T> coll, T... ts)
```

This method can now be called with generic types. You can use this annotation for any methods that merely read the elements of the parameter array, which is bound to be the most common use case.

The @SafeVarargs can only be used with constructors and methods that are static, final, or (as of Java 9) private. Any other method could be overridden, making the annotation meaningless.

 NOTE: You can use the @SafeVarargs annotation to defeat the restriction against generic array creation, using this method:

```
@SafeVarargs static <E> E[] array(E... array) { return array; }
```

Now you can call

```
Pair<String>[] table = array(pair1, pair2);
```

This seems convenient, but there is a hidden danger. The code

```
Object[] objarray = table;
objarray[0] = new Pair<Employee>();
```

will run without an ArrayStoreException (because the array store only checks the erased type), and you'll get an exception elsewhere when you work with table[0].

8.6.5 You Cannot Instantiate Type Variables

You cannot use type variables in an expression such as new T(. . .). For example, the following Pair<T> constructor is illegal:

```
public Pair() { first = new T(); second = new T(); } // ERROR
```

Type erasure would change T to Object, and surely you don't want to call new Object().

The best workaround, available since Java 8, is to make the caller provide a constructor expression. For example:

```
Pair<String> p = Pair.makePair(String::new);
```

The makePair method receives a Supplier<T>, the functional interface for a function with no arguments and a result of type T:

```
public static <T> Pair<T> makePair(Supplier<T> constr)
{
    return new Pair<>(constr.get(), constr.get());
}
```

A more traditional workaround is to construct generic objects through reflection, by calling the Constructor.newInstance method.

Unfortunately, the details are a bit complex. You cannot call

```
first = T.class.getConstructor().newInstance(); // ERROR
```

The expression T.class is not legal because it would erase to Object.class. Instead, you must design the API so that you are handed a Class object, like this:

```
public static <T> Pair<T> makePair(Class<T> cl)
{
    try {
        return new Pair<>(cl.getConstructor().newInstance(),
            cl.getConstructor().newInstance());
    }
    catch (Exception e) { return null; }
}
```

This method could be called as follows:

```
Pair<String> p = Pair.makePair(String.class);
```

Note that the Class class is itself generic. For example, String.class is an instance (indeed, the sole instance) of Class<String>. Therefore, the makePair method can infer the type of the pair that it is making.

8.6.6 You Cannot Construct a Generic Array

Just as you cannot instantiate a single generic instance, you cannot instantiate an array. The reasons are different—an array is, after all, filled with null values, which would seem safe to construct. But an array also carries a type, which is used to monitor array stores in the virtual machine. That type is erased. For example, consider

```
public static <T extends Comparable> T[] minmax(T... a)
{
   T[] mm = new T[2]; // ERROR
   . . .
}
```

Type erasure would cause this method to always construct an array Comparable[2].

If the array is only used as a private instance field of a class, you can declare the element type of the array to be the erased type and use casts. For example, the ArrayList class could be implemented as follows:

```
public class ArrayList<E>
{
   private Object[] elements;
   . . .
   @SuppressWarnings("unchecked") public E get(int n) { return (E) elements[n]; }
   public void set(int n, E e) { elements[n] = e; } // no cast needed
}
```

The actual implementation is not quite as clean:

```
public class ArrayList<E>
{
   private E[] elements;
   . . .
   public ArrayList() { elements = (E[]) new Object[10]; }
}
```

Here, the cast E[] is an outright lie, but type erasure makes it undetectable.

This technique does not work for our minmax method since we are returning a T[] array, and a runtime error results if we lie about its type. Suppose we implement

```
public static <T extends Comparable> T[] minmax(T... a)
{
   var result = new Comparable[2]; // array of erased type
   . . .
   return (T[]) result; // compiles with warning
}
```

The call

```
String[] names = ArrayAlg.minmax("Tom", "Dick", "Harry");
```

compiles without any warning. A ClassCastException occurs when the Comparable[] reference is cast to String[] after the method returns.

In this situation, it is best to ask the user to provide an array constructor expression:

```
String[] names = ArrayAlg.minmax(String[]::new, "Tom", "Dick", "Harry");
```

The constructor expression String[]::new denotes a function that, given the desired length, constructs a String array of that length.

The method uses that parameter to produce an array of the correct type:

```
public static <T extends Comparable> T[] minmax(IntFunction<T[]> constr, T... a)
{
   T[] result = constr.apply(2);
   . . .
}
```

A more old-fashioned approach is to use reflection and call Array.newInstance:

```
public static <T extends Comparable> T[] minmax(T... a)
{
   var result = (T[]) Array.newInstance(a.getClass().getComponentType(), 2);
   . . .
}
```

The toArray method of the ArrayList class is not so lucky. It needs to produce a T[] array, but it doesn't have the component type. Therefore, there are two variants:

```
Object[] toArray()
T[] toArray(T[] result)
```

The second method receives an array parameter. If the array is large enough, it is used. Otherwise, a new array of sufficient size is created, using the component type of result.

8.6.7 Type Variables Are Not Valid in Static Contexts of Generic Classes

You cannot reference type variables in static fields or methods. For example, the following clever idea won't work:

```
public class Singleton<T>
{
   private static T singleInstance; // ERROR

   public static T getSingleInstance() // ERROR
   {
```

```
      if (singleInstance == null) construct new instance of T
      return singleInstance;
   }
}
```

If this could be done, then a program could declare a Singleton<Random> to share a random number generator and a Singleton<JFileChooser> to share a file chooser dialog. But it can't work. After type erasure there is only one Singleton class, and only one singleInstance field. For that reason, static fields and methods with type variables are simply outlawed.

8.6.8 You Cannot Throw or Catch Instances of a Generic Class

You can neither throw nor catch objects of a generic class. In fact, it is not even legal for a generic class to extend Throwable. For example, the following definition will not compile:

```
public class Problem<T> extends Exception { /* . . . */ }
   // ERROR--can't extend Throwable
```

You cannot use a type variable in a catch clause. For example, the following method will not compile:

```
public static <T extends Throwable> void doWork(Class<T> t)
{
   try
   {
      do work
   }
   catch (T e) // ERROR--can't catch type variable
   {
      Logger.global.info(. . .);
   }
}
```

However, it is OK to use type variables in exception specifications. The following method is legal:

```
public static <T extends Throwable> void d
   {
      do work
   }
   catch (Throwable realCause)
   {
      t.initCause(realCause);
      throw t;
   }
}
```

8.6.9 You Can Defeat Checked Exception Checking

A bedrock principle of Java exception handling is that you must provide a handler for all checked exceptions. You can use generics to defeat this scheme. The key ingredient is this method:

```
@SuppressWarnings("unchecked")
static <T extends Throwable> void throwAs(Throwable t) throws T
{
   throw (T) t;
}
```

Suppose this method is contained in an interface Task. When you have a checked exception e and call

```
Task.<RuntimeException>throwAs(e);
```

then the compiler will believe that e becomes an unchecked exception. The following turns all exceptions into those that the compiler believes to be unchecked:

```
try
{
   do work
}
catch (Throwable t)
{
   Task.<RuntimeException>throwAs(t);
}
```

Let's use this to solve a vexing problem. To run code in a thread, you have to place it into the run method of a class that implements the Runnable interface. But that method is not allowed to throw checked exceptions. We will provide an adaptor from a Task, whose run method is allowed to throw arbitrary exceptions, to a Runnable:

```
interface Task
{
   void run() throws Exception;

   @SuppressWarnings("unchecked")
   static <T extends Throwable> void throwAs(Throwable t) throws T
   {
      throw (T) t;
   }

   static Runnable asRunnable(Task task)
   {
      return () ->
      {
```

```
        try
        {
           task.run();
        }
        catch (Exception e)
        {
           Task.<RuntimeException>throwAs(e);
        }
     };
  }
}
```

For example, this program runs a thread that will throw a checked exception:

```
public class Test
{
   public static void main(String[] args)
   {
      var thread = new Thread(Task.asRunnable(() ->
         {
            Thread.sleep(1000);
            System.out.println("Hello, World!");
            throw new Exception("Check this out!");
         }));
      thread.start();
   }
}
```

The `Thread.sleep` method is declared to throw an `InterruptedException`, and we no longer have to catch it. Since we don't interrupt the thread, that exception won't be thrown. However, the program throws a checked exception. When you run the program, you will get a stack trace.

What's so remarkable about that? Normally, you have to catch all checked exceptions inside the `run` method of a `Runnable` and *wrap them* into unchecked exceptions—the `run` method is declared to throw no checked exceptions.

But here, we don't wrap. We simply throw the exception, tricking the compiler into believing that it is not a checked exception.

Using generic classes, erasure, and the `@SuppressWarnings` annotation, we were able to defeat an essential part of the Java type system.

8.6.10 Beware of Clashes after Erasure

It is illegal to create conditions that cause clashes when generic types are erased. Here is an example. Suppose we add an `equals` method to the `Pair` class, like this:

```
public class Pair<T>
{
   public boolean equals(T value) { return first.equals(value) && second.equals(value); }
   . . .
}
```

Consider a Pair<String>. Conceptually, it has two equals methods:

```
boolean equals(String) // defined in Pair<T>
boolean equals(Object) // inherited from Object
```

But the intuition leads us astray. The erasure of the method

```
boolean equals(T)
```

is

```
boolean equals(Object)
```

which clashes with the Object.equals method.

The remedy is, of course, to rename the offending method.

The generics specification cites another rule: "To support translation by erasure, we impose the restriction that a class or type variable may not at the same time be a subtype of two interface types which are different parameterizations of the same interface." For example, the following is illegal:

```
class Employee implements Comparable<Employee> { . . . }
class Manager extends Employee implements Comparable<Manager> { . . . } // ERROR
```

Manager would then implement both Comparable<Employee> and Comparable<Manager>, which are different parameterizations of the same interface.

It is not obvious what this restriction has to do with type erasure. After all, the nongeneric version

```
class Employee implements Comparable { . . . }
class Manager extends Employee implements Comparable { . . . }
```

is legal. The reason is far more subtle. There would be a conflict with the synthesized bridge methods. A class that implements Comparable<X> gets a bridge method

```
public int compareTo(Object other) { return compareTo((X) other); }
```

You cannot have two such methods for different types X.

8.7 Inheritance Rules for Generic Types

When you work with generic classes, you need to learn a few rules about inheritance and subtypes. Let's start with a situation which many programmers find unintuitive. Consider a class and a subclass, such as Employee and Manager. Is Pair<Manager> a subclass of Pair<Employee>? Perhaps surprisingly, the answer is "no." For example, the following code will not compile:

```
Manager[] topHonchos = . . .;
Pair<Employee> result = ArrayAlg.minmax(topHonchos); // ERROR
```

The minmax method returns a Pair<Manager>, not a Pair<Employee>, and it is illegal to assign one to the other.

In general, there is *no* relationship between Pair<S> and Pair<T>, no matter how S and T are related (see Figure 8.1).

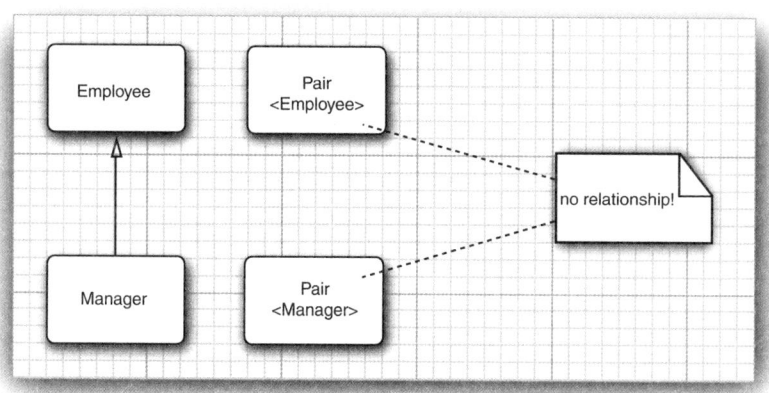

Figure 8.1 No inheritance relationship between pair classes

This seems like a cruel restriction, but it is necessary for type safety. Suppose we were allowed to convert a Pair<Manager> to a Pair<Employee>. Consider this code:

```
var managerBuddies = new Pair<Manager>(ceo, cfo);
Pair<Employee> employeeBuddies = managerBuddies; // illegal, but suppose it wasn't
employeeBuddies.setFirst(lowlyEmployee);
```

Clearly, the last statement is legal. But employeeBuddies and managerBuddies refer to the *same object*. We now managed to pair up the CFO with a lowly employee, which should not be possible for a Pair<Manager>.

NOTE: You just saw an important difference between generic types and Java arrays. You can assign a `Manager[]` array to a variable of type `Employee[]`:

```
Manager[] managerBuddies = { ceo, cfo };
Employee[] employeeBuddies = managerBuddies; // OK
```

However, arrays come with special protection. If you try to store a lowly employee into `employeeBuddies[0]`, the virtual machine throws an `ArrayStoreException`.

You can always convert a parameterized type to a raw type. For example, `Pair<Employee>` is a subtype of the raw type `Pair`. This conversion is necessary for interfacing with legacy code.

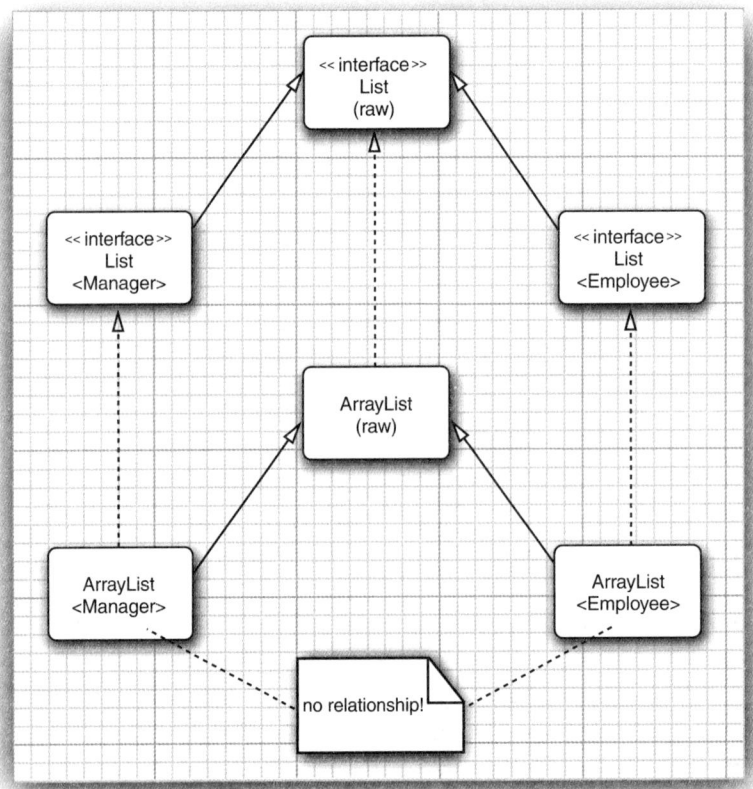

Figure 8.2 Subtype relationships among generic list types

Can you convert to the raw type and then cause a type error? Unfortunately, you can. Consider this example:

```
var managerBuddies = new Pair<Manager>(ceo, cfo);
Pair rawBuddies = managerBuddies; // OK
rawBuddies.setFirst(new File(". . .")); // only a compile-time warning
```

This sounds scary. However, keep in mind that you are no worse off than you were with older versions of Java. The security of the virtual machine is not at stake. When the foreign object is retrieved with getFirst and assigned to a Manager variable, a ClassCastException is thrown, just as in the good old days. You merely lose the added safety that generic programming normally provides.

Finally, generic classes can extend or implement other generic classes. In this regard, they are no different from ordinary classes. For example, the class ArrayList<T> implements the interface List<T>. That means an ArrayList<Manager> can be converted to a List<Manager>. However, as you just saw, an ArrayList<Manager> is *not* an ArrayList<Employee> or List<Employee>. Figure 8.2 shows these relationships.

8.8 Wildcard Types

It was known for some time among researchers of type systems that a rigid system of generic types is quite unpleasant to use. The Java designers invented an ingenious (but nevertheless safe) "escape hatch": the *wildcard type*. The following sections show you how to work with wildcards.

8.8.1 The Wildcard Concept

In a wildcard type, a type parameter is allowed to vary. For example, the wildcard type

```
Pair<? extends Employee>
```

denotes any generic Pair type whose type parameter is a subclass of Employee, such as Pair<Manager>, but not Pair<String>.

Let's say you want to write a method that prints out pairs of employees, like this:

```
public static void printBuddies(Pair<Employee> p)
{
   Employee first = p.getFirst();
   Employee second = p.getSecond();
   System.out.println(first.getName() + " and " + second.getName() + " are buddies.");
}
```

As you saw in the preceding section, you cannot pass a Pair<Manager> to that method, which is rather limiting. But the solution is simple—use a wildcard type:

```
public static void printBuddies(Pair<? extends Employee> p)
```

The type Pair<Manager> is a subtype of Pair<? extends Employee> (see Figure 8.3).

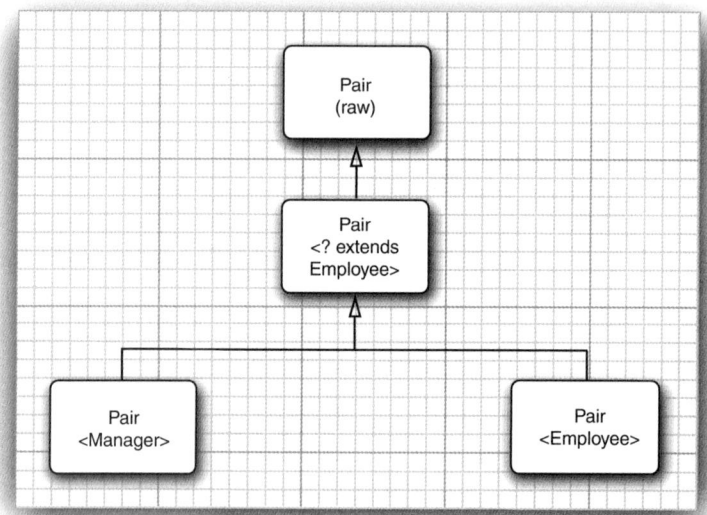

Figure 8.3 Subtype relationships with wildcards

Can we use wildcards to corrupt a Pair<Manager> through a Pair<? extends Employee> reference?

```
var managerBuddies = new Pair<Manager>(ceo, cfo);
Pair<? extends Employee> wildcardBuddies = managerBuddies; // OK
wildcardBuddies.setFirst(lowlyEmployee); // compile-time error
```

No corruption is possible. The call to setFirst is a type error. To see why, let us have a closer look at the type Pair<? extends Employee>. Its methods look like this:

```
? extends Employee getFirst()
void setFirst(? extends Employee)
```

This makes it impossible to call the setFirst method. The compiler only knows that it needs some subtype of Employee, but it doesn't know which type. It refuses to pass any specific type—after all, ? might not match it.

We don't have this problem with getFirst: It is perfectly legal to assign the return value of getFirst to an Employee reference.

This is the key idea behind bounded wildcards. We now have a way of distinguishing between the safe accessor methods and the unsafe mutator methods.

8.8.2 Supertype Bounds for Wildcards

Wildcard bounds are similar to type variable bounds, but they have an added capability—you can specify a *supertype bound*, like this:

```
? super Manager
```

This wildcard is restricted to all supertypes of Manager. (It was a stroke of good luck that the existing super keyword describes the relationship so accurately.)

Why would you want to do this? A wildcard with a supertype bound gives you a behavior that is opposite to that of the wildcards described in Section 8.8, "Wildcard Types," on p. 459. You can supply parameters to methods, but you can't use the return values. For example, Pair<? super Manager> has methods that can be described as follows:

```
void setFirst(? super Manager)
? super Manager getFirst()
```

This is not actual Java syntax, but it shows what the compiler knows. The compiler cannot know the exact type of the setFirst method and therefore cannot accept a call with an argument of type Employee or Object. It is only possible to pass an object of type Manager or a subtype such as Executive. Moreover, if you call getFirst, there is no guarantee about the type of the returned object. You can only assign it to an Object.

Here is a typical example. We have an array of managers and want to put the manager with the lowest and highest bonus into a Pair object. What kind of Pair? A Pair<Employee> should be fair game or, for that matter, a Pair<Object> (see Figure 8.4). The following method will accept any appropriate Pair:

```
public static void minmaxBonus(Manager[] a, Pair<? super Manager> result)
{
   if (a.length == 0) return;
   Manager min = a[0];
   Manager max = a[0];
   for (int i = 1; i < a.length; i++)
   {
      if (min.getBonus() > a[i].getBonus()) min = a[i];
      if (max.getBonus() < a[i].getBonus()) max = a[i];
   }
```

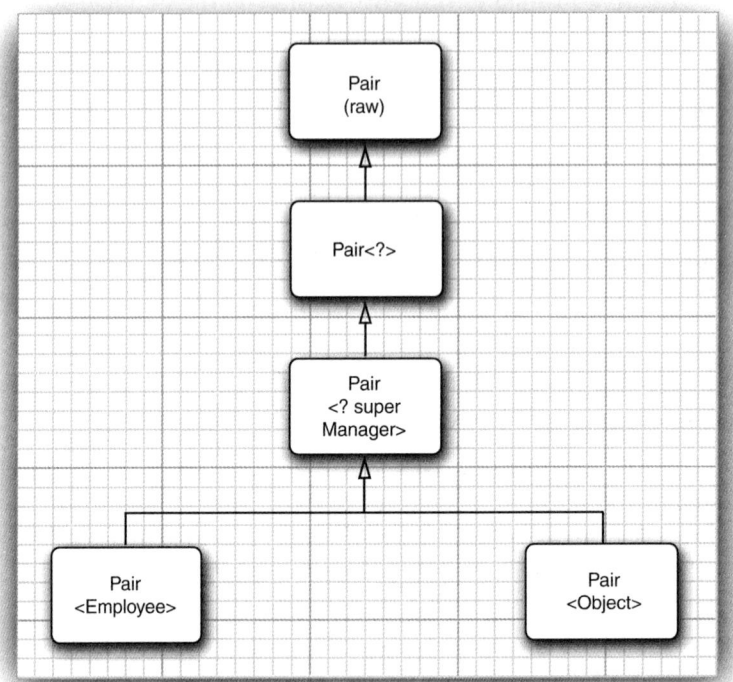

Figure 8.4 A wildcard with a supertype bound

```
      result.setFirst(min);
      result.setSecond(max);
   }
```

Intuitively speaking, wildcards with supertype bounds let you write to a generic object, while wildcards with subtype bounds let you read from a generic object.

Here is another use for supertype bounds. The Comparable interface is itself a generic type. It is declared as follows:

```
public interface Comparable<T>
{
   public int compareTo(T other);
}
```

Here, the type variable indicates the type of the `other` parameter. For example, the `String` class implements `Comparable<String>`, and its `compareTo` method is declared as

```
public int compareTo(String other)
```

This is nice—the explicit parameter has the correct type. Before the interface was generic, `other` was an `Object`, and a cast was necessary in the implementation of the method.

Now that `Comparable` is a generic type, perhaps we should have done a better job with the `min` method of the `ArrayAlg` class? We could have declared it as

```
public static <T extends Comparable<T>> T min(T[] a)
```

This looks more thorough than just using `T extends Comparable`, and it would work fine for many classes. For example, if you compute the minimum of a `String` array, then `T` is the type `String`, and `String` is a subtype of `Comparable<String>`. But we run into a problem when processing an array of `LocalDate` objects. As it happens, `LocalDate` implements `ChronoLocalDate`, and `ChronoLocalDate` extends `Comparable<ChronoLocalDate>`. Thus, `LocalDate` implements `Comparable<ChronoLocalDate>` but *not* `Comparable<LocalDate>`.

In a situation such as this one, supertypes come to the rescue:

```
public static <T extends Comparable<? super T>> T min(T[] a) . . .
```

Now the `compareTo` method has the form

```
int compareTo(? super T)
```

Maybe it is declared to take an object of type `T`, or—for example, when `T` is `LocalDate`—a supertype of `T`. At any rate, it is safe to pass an object of type `T` to the `compareTo` method.

To the uninitiated, a declaration such as `<T extends Comparable<? super T>>` is bound to look intimidating. This is unfortunate, because the intent of this declaration is to help application programmers by removing unnecessary restrictions on the call parameters. Application programmers with no interest in generics will probably learn quickly to gloss over these declarations and just take for granted that library programmers will do the right thing. If you are a library programmer, you'll need to get used to wildcards, or your users will curse you and throw random casts at their code until it compiles.

NOTE: Another common use for supertype bounds is an argument type of a functional interface. For example, the `Collection` interface has a method

```
default boolean removeIf(Predicate<? super E> filter)
```

The method removes all elements that fulfill the given predicate. For example, if you hate employees with odd hash codes, you can remove them like this:

```
ArrayList<Employee> staff = . . .;
Predicate<Object> oddHashCode = obj -> obj.hashCode() %2 != 0;
staff.removeIf(oddHashCode);
```

You want to be able to pass a `Predicate<Object>`, not just a `Predicate<Employee>`. The `super` wildcard makes that possible.

8.8.3 Unbounded Wildcards

You can even use wildcards with no bounds at all—for example, `Pair<?>`. At first glance, this looks identical to the raw `Pair` type. Actually, the types are very different. The type `Pair<?>` has methods such as

```
? getFirst()
void setFirst(?)
```

The return value of `getFirst` can only be assigned to an `Object`. The `setFirst` method can never be called, *not even with an* `Object`. That's the essential difference between `Pair<?>` and `Pair`: you can call the `setFirst` method of the raw `Pair` class with *any* `Object`.

 NOTE: You can call `setFirst(null)`.

Why would you ever want such a wimpy type? It is useful for very simple operations. For example, the following method tests whether a pair contains a `null` reference. It never needs the actual type.

```
public static boolean hasNulls(Pair<?> p)
{
   return p.getFirst() == null || p.getSecond() == null;
}
```

You could have avoided the wildcard type by turning `hasNulls` into a generic method:

```
public static <T> boolean hasNulls(Pair<T> p)
```

However, the version with the wildcard type seems easier to read.

8.8.4 Wildcard Capture

Let us write a method that swaps the elements of a pair:

```
public static void swap(Pair<?> p)
```

A wildcard is not a type variable, so we can't write code that uses ? as a type. In other words, the following would be illegal:

```
? t = p.getFirst(); // ERROR
p.setFirst(p.getSecond());
p.setSecond(t);
```

That's a problem because we need to temporarily hold the first element when we do the swapping. Fortunately, there is an interesting solution to this problem. We can write a helper method, swapHelper, like this:

```
public static <T> void swapHelper(Pair<T> p)
{
   T t = p.getFirst();
   p.setFirst(p.getSecond());
   p.setSecond(t);
}
```

Note that swapHelper is a generic method, whereas swap is not—it has a fixed parameter of type Pair<?>.

Now we can call swapHelper from swap:

```
public static void swap(Pair<?> p) { swapHelper(p); }
```

In this case, the parameter T of the swapHelper method *captures the wildcard*. It isn't known what type the wildcard denotes, but it is a definite type, and the definition of <T>swapHelper makes perfect sense when T denotes that type.

Of course, in this case, we were not compelled to use a wildcard. We could have directly implemented <T> void swap(Pair<T> p) as a generic method without wildcards. However, consider this example in which a wildcard type occurs naturally in the middle of a computation:

```
public static void maxminBonus(Manager[] a, Pair<? super Manager> result)
{
   minmaxBonus(a, result);
   PairAlg.swapHelper(result); // OK--swapHelper captures wildcard type
}
```

Here, the wildcard capture mechanism cannot be avoided.

Wildcard capture is only legal in very limited circumstances. The compiler must be able to guarantee that the wildcard represents a single, definite type. For example, the T in ArrayList<Pair<T>> can never capture the wildcard in

ArrayList<Pair<?>>. The array list might hold two Pair<?>, each of which has a different type for ?.

The test program in Listing 8.3 gathers up the various methods that we discussed in the preceding sections so you can see them in context.

Listing 8.3 pair3/PairTest3.java

```
1  package pair3;
2
3  /**
4   * @version 1.01 2012-01-26
5   * @author Cay Horstmann
6   */
7  public class PairTest3
8  {
9     public static void main(String[] args)
10    {
11       var ceo = new Manager("Gus Greedy", 800000, 2003, 12, 15);
12       var cfo = new Manager("Sid Sneaky", 600000, 2003, 12, 15);
13       var buddies = new Pair<Manager>(ceo, cfo);
14       printBuddies(buddies);
15
16       ceo.setBonus(1000000);
17       cfo.setBonus(500000);
18       Manager[] managers = { ceo, cfo };
19
20       var result = new Pair<Employee>();
21       minmaxBonus(managers, result);
22       System.out.println("first: " + result.getFirst().getName()
23          + ", second: " + result.getSecond().getName());
24       maxminBonus(managers, result);
25       System.out.println("first: " + result.getFirst().getName()
26          + ", second: " + result.getSecond().getName());
27    }
28
29    public static void printBuddies(Pair<? extends Employee> p)
30    {
31       Employee first = p.getFirst();
32       Employee second = p.getSecond();
33       System.out.println(first.getName() + " and " + second.getName() + " are buddies.");
34    }
35
36    public static void minmaxBonus(Manager[] a, Pair<? super Manager> result)
37    {
38       if (a.length == 0) return;
39       Manager min = a[0];
40       Manager max = a[0];
```

```
41      for (int i = 1; i < a.length; i++)
42      {
43         if (min.getBonus() > a[i].getBonus()) min = a[i];
44         if (max.getBonus() < a[i].getBonus()) max = a[i];
45      }
46      result.setFirst(min);
47      result.setSecond(max);
48   }
49
50   public static void maxminBonus(Manager[] a, Pair<? super Manager> result)
51   {
52      minmaxBonus(a, result);
53      PairAlg.swapHelper(result); // OK--swapHelper captures wildcard type
54   }
55   // can't write public static <T super manager> . . .
56 }
57
58 class PairAlg
59 {
60   public static boolean hasNulls(Pair<?> p)
61   {
62      return p.getFirst() == null || p.getSecond() == null;
63   }
64
65   public static void swap(Pair<?> p) { swapHelper(p); }
66
67   public static <T> void swapHelper(Pair<T> p)
68   {
69      T t = p.getFirst();
70      p.setFirst(p.getSecond());
71      p.setSecond(t);
72   }
73 }
```

8.9 Reflection and Generics

Reflection lets you analyze arbitrary objects at runtime. If the objects are instances of generic classes, you don't get much information about the generic type parameters because they have been erased. In the following sections, you will learn what you can nevertheless find out about generic classes with reflection.

8.9.1 The Generic Class Class

The Class class is now generic. For example, String.class is actually an object (in fact, the sole object) of the class Class<String>.

The type parameter is useful because it allows the methods of Class<T> to be more specific about their return types. The following methods of Class<T> take advantage of the type parameter:

```
T newInstance()
T cast(Object obj)
T[] getEnumConstants()
Class<? super T> getSuperclass()
Constructor<T> getConstructor(Class... parameterTypes)
Constructor<T> getDeclaredConstructor(Class... parameterTypes)
```

The newInstance method returns an instance of the class, obtained from the no-argument constructor. Its return type can now be declared to be T, the same type as the class that is being described by Class<T>. That saves a cast.

The cast method returns the given object, now declared as type T if its type is indeed a subtype of T. Otherwise, it throws a BadCastException.

The getEnumConstants method returns null if this class is not an enum class or an array of the enumeration values which are known to be of type T.

Finally, the getConstructor and getDeclaredConstructor methods return a Constructor<T> object. The Constructor class has also been made generic so that its newInstance method has the correct return type.

---

java.lang.Class<T>   1.0

- T newInstance()

  returns a new instance constructed with the no-argument constructor.

- T cast(Object obj)

  returns obj if it is null or can be converted to the type T, or throws a BadCastException otherwise.

- T[] getEnumConstants() 5

  returns an array of all values if T is an enumerated type, null otherwise.

- Class<? super T> getSuperclass()

  returns the superclass of this class, or null if T is not a class or the class Object.

- Constructor<T> getConstructor(Class... parameterTypes) 1.1
- Constructor<T> getDeclaredConstructor(Class... parameterTypes) 1.1

  gets the public constructor, or the constructor with the given parameter types.

```
java.lang.reflect.Constructor<T>  1.1
```

• T newInstance(Object... parameters)

   returns a new instance constructed with the given parameters.

8.9.2 Using Class<T> Parameters for Type Matching

It is sometimes useful to match the type variable of a Class<T> parameter in a generic method. Here is the canonical example:

```
public static <T> Pair<T> makePair(Class<T> c) throws InstantiationException,
    IllegalAccessException
{
    return new Pair<>(c.newInstance(), c.newInstance());
}
```

If you call

```
makePair(Employee.class)
```

then Employee.class is an object of type Class<Employee>. The type parameter T of the makePair method matches Employee, and the compiler can infer that the method returns a Pair<Employee>.

8.9.3 Generic Type Information in the Virtual Machine

One of the notable features of Java generics is the erasure of generic types in the virtual machine. Perhaps surprisingly, the erased classes still retain some faint memory of their generic origin. For example, the raw Pair class knows that it originated from the generic class Pair<T>, even though an object of type Pair can't tell whether it was constructed as a Pair<String> or Pair<Employee>.

Similarly, consider a method

```
public static Comparable min(Comparable[] a)
```

that is the erasure of a generic method

```
public static <T extends Comparable<? super T>> T min(T[] a)
```

You can use the reflection API to determine that

• The generic method has a type parameter called T;
• The type parameter has a subtype bound that is itself a generic type;
• The bounding type has a wildcard parameter;
• The wildcard parameter has a supertype bound; and
• The generic method has a generic array parameter.

In other words, you can reconstruct everything about generic classes and methods that their implementors declared. However, you won't know how the type parameters were resolved for specific objects or method calls.

In order to express generic type declarations, use the interface Type in the java.lang.reflect package. The interface has the following subtypes:

- The Class class, describing concrete types
- The TypeVariable interface, describing type variables (such as T extends Comparable<? super T>)
- The WildcardType interface, describing wildcards (such as ? super T)
- The ParameterizedType interface, describing generic class or interface types (such as Comparable<? super T>)
- The GenericArrayType interface, describing generic arrays (such as T[])

Figure 8.5 shows the inheritance hierarchy. Note that the last four subtypes are interfaces—the virtual machine instantiates suitable classes that implement these interfaces.

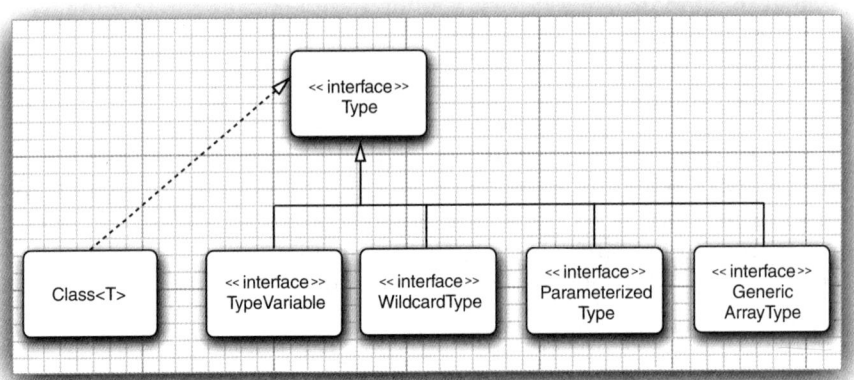

Figure 8.5 The Type interface and its descendants

Listing 8.4 uses the generic reflection API to print out what it discovers about a given class. If you run it with the Pair class, you get this report:

```
class Pair<T> extends java.lang.Object
public T getFirst()
public T getSecond()
public void setFirst(T)
public void setSecond(T)
```

If you run it with `ArrayAlg` in the `PairTest2` directory, the report displays the following method:

```
public static <T extends java.lang.Comparable> Pair<T> minmax(T[])
```

Listing 8.4 genericReflection/GenericReflectionTest.java

```
1  package genericReflection;
2
3  import java.lang.reflect.*;
4  import java.util.*;
5
6  /**
7   * @version 1.11 2018-04-10
8   * @author Cay Horstmann
9   */
10 public class GenericReflectionTest
11 {
12    public static void main(String[] args)
13    {
14       // read class name from command line args or user input
15       String name;
16       if (args.length > 0) name = args[0];
17       else
18       {
19          try (var in = new Scanner(System.in))
20          {
21             System.out.println("Enter class name (e.g., java.util.Collections): ");
22             name = in.next();
23          }
24       }
25
26       try
27       {
28          // print generic info for class and public methods
29          Class<?> cl = Class.forName(name);
30          printClass(cl);
31          for (Method m : cl.getDeclaredMethods())
32             printMethod(m);
33       }
34       catch (ClassNotFoundException e)
35       {
36          e.printStackTrace();
37       }
38    }
39
40    public static void printClass(Class<?> cl)
41    {
```

(Continues)

Listing 8.4 *(Continued)*

```
42       System.out.print(cl);
43       printTypes(cl.getTypeParameters(), "<", ", ", ">", true);
44       Type sc = cl.getGenericSuperclass();
45       if (sc != null)
46       {
47          System.out.print(" extends ");
48          printType(sc, false);
49       }
50       printTypes(cl.getGenericInterfaces(), " implements ", ", ", "", false);
51       System.out.println();
52    }
53
54    public static void printMethod(Method m)
55    {
56       String name = m.getName();
57       System.out.print(Modifier.toString(m.getModifiers()));
58       System.out.print(" ");
59       printTypes(m.getTypeParameters(), "<", ", ", "> ", true);
60
61       printType(m.getGenericReturnType(), false);
62       System.out.print(" ");
63       System.out.print(name);
64       System.out.print("(");
65       printTypes(m.getGenericParameterTypes(), "", ", ", "", false);
66       System.out.println(")");
67    }
68
69    public static void printTypes(Type[] types, String pre, String sep, String suf,
70          boolean isDefinition)
71    {
72       if (pre.equals(" extends ") && Arrays.equals(types, new Type[] { Object.class }))
73          return;
74       if (types.length > 0) System.out.print(pre);
75       for (int i = 0; i < types.length; i++)
76       {
77          if (i > 0) System.out.print(sep);
78          printType(types[i], isDefinition);
79       }
80       if (types.length > 0) System.out.print(suf);
81    }
82
83    public static void printType(Type type, boolean isDefinition)
84    {
85       if (type instanceof Class)
86       {
```

```
87           var t = (Class<?>) type;
88           System.out.print(t.getName());
89         }
90         else if (type instanceof TypeVariable)
91         {
92           var t = (TypeVariable<?>) type;
93           System.out.print(t.getName());
94           if (isDefinition)
95             printTypes(t.getBounds(), " extends ", " & ", "", false);
96         }
97         else if (type instanceof WildcardType)
98         {
99           var t = (WildcardType) type;
100          System.out.print("?");
101          printTypes(t.getUpperBounds(), " extends ", " & ", "", false);
102          printTypes(t.getLowerBounds(), " super ", " & ", "", false);
103        }
104        else if (type instanceof ParameterizedType)
105        {
106          var t = (ParameterizedType) type;
107          Type owner = t.getOwnerType();
108          if (owner != null)
109          {
110            printType(owner, false);
111            System.out.print(".");
112          }
113          printType(t.getRawType(), false);
114          printTypes(t.getActualTypeArguments(), "<", ", ", ">", false);
115        }
116        else if (type instanceof GenericArrayType)
117        {
118          var t = (GenericArrayType) type;
119          System.out.print("");
120          printType(t.getGenericComponentType(), isDefinition);
121          System.out.print("[]");
122        }
123      }
124 }
```

8.9.4 Type Literals

Sometimes, you want to drive program behavior by the type of a value. For example, in a persistence mechanism, you may want the user to specify a way of saving an object of a particular class. This is typically implemented by associating the Class object with an action.

However, with generic classes, erasure poses a problem. How can you have different actions for, say, ArrayList<Integer> and ArrayList<String> when both erase to the same raw ArrayList type?

There is a trick that can offer relief in some situations. You can capture an instance of the `Type` interface that you encountered in the preceding section. Construct an anonymous subclass like this:

```
var type = new TypeLiteral<ArrayList<Integer>>(){} // note the {}
```

The `TypeLiteral` constructor captures the generic supertype:

```
class TypeLiteral
{
    public TypeLiteral()
    {
        Type parentType = getClass().getGenericSuperclass();
        if (parentType instanceof ParameterizedType)
        {
            type = ((ParameterizedType) parentType).getActualTypeArguments()[0];
        }
        else
            throw new UnsupportedOperationException(
                "Construct as new TypeLiteral<. . .>(){}");
    }
    . . .
}
```

If we have a generic type available at runtime, we can match it against the `TypeLiteral`. We can't get a generic type from an object—it is erased. But, as you have seen in the preceding section, generic types of fields and method parameters survive in the virtual machine.

Injection frameworks such as CDI and Guice use type literals to control injection of generic types. The example program in the book's companion code shows a simpler example. Given an object, we enumerate its fields, whose generic types are available, and look up associated formatting actions.

We format an `ArrayList<Integer>` by separating the values with spaces, an `ArrayList<Character>` by joining the characters to a string. Any other array lists are formatted by `ArrayList.toString`.

Listing 8.5 genericReflection/TypeLiterals.java

```
1  package genericReflection;
2
3  /**
4     @version 1.01 2018-04-10
5     @author Cay Horstmann
6  */
7
8  import java.lang.reflect.*;
9  import java.util.*;
```

```
10  import java.util.function.*;
11
12  /**
13   * A type literal describes a type that can be generic, such as ArrayList<String>.
14   */
15  class TypeLiteral<T>
16  {
17     private Type type;
18
19     /**
20      * This constructor must be invoked from an anonymous subclass
21      * as new TypeLiteral<. . .>(){}
22      */
23     public TypeLiteral()
24     {
25        Type parentType = getClass().getGenericSuperclass();
26        if (parentType instanceof ParameterizedType)
27        {
28           type = ((ParameterizedType) parentType).getActualTypeArguments()[0];
29        }
30        else
31           throw new UnsupportedOperationException(
32              "Construct as new TypeLiteral<. . .>(){}");
33     }
34
35     private TypeLiteral(Type type)
36     {
37        this.type = type;
38     }
39
40     /**
41      * Yields a type literal that describes the given type.
42      */
43     public static TypeLiteral<?> of(Type type)
44     {
45        return new TypeLiteral<Object>(type);
46     }
47
48     public String toString()
49     {
50        if (type instanceof Class) return ((Class<?>) type).getName();
51        else return type.toString();
52     }
53
54     public boolean equals(Object otherObject)
55     {
56        return otherObject instanceof TypeLiteral
57           && type.equals(((TypeLiteral<?>) otherObject).type);
58     }
```

(Continues)

Listing 8.5 *(Continued)*

```
59
60     public int hashCode()
61     {
62        return type.hashCode();
63     }
64  }
65
66  /**
67   * Formats objects, using rules that associate types with formatting functions.
68   */
69  class Formatter
70  {
71     private Map<TypeLiteral<?>, Function<?, String>> rules = new HashMap<>();
72
73     /**
74      * Add a formatting rule to this formatter.
75      * @param type the type to which this rule applies
76      * @param formatterForType the function that formats objects of this type
77      */
78     public <T> void forType(TypeLiteral<T> type, Function<T, String> formatterForType)
79     {
80        rules.put(type,  formatterForType);
81     }
82
83     /**
84      * Formats all fields of an object using the rules of this formatter.
85      * @param obj an object
86      * @return a string with all field names and formatted values
87      */
88     public String formatFields(Object obj)
89           throws IllegalArgumentException, IllegalAccessException
90     {
91        var result = new StringBuilder();
92        for (Field f : obj.getClass().getDeclaredFields())
93        {
94           result.append(f.getName());
95           result.append("=");
96           f.setAccessible(true);
97           Function<?, String> formatterForType = rules.get(TypeLiteral.of(f.getGenericType()));
98           if (formatterForType != null)
99           {
100             // formatterForType has parameter type ?. Nothing can be passed to its apply
101             // method. Cast makes the parameter type to Object so we can invoke it.
102             @SuppressWarnings("unchecked")
103             Function<Object, String> objectFormatter
104                = (Function<Object, String>) formatterForType;
```

```
105          result.append(objectFormatter.apply(f.get(obj)));
106       }
107       else
108          result.append(f.get(obj).toString());
109       result.append("\n");
110    }
111    return result.toString();
112    }
113 }
114
115 public class TypeLiterals
116 {
117    public static class Sample
118    {
119       ArrayList<Integer> nums;
120       ArrayList<Character> chars;
121       ArrayList<String> strings;
122       public Sample()
123       {
124          nums = new ArrayList<>();
125          nums.add(42); nums.add(1729);
126          chars = new ArrayList<>();
127          chars.add('H'); chars.add('i');
128          strings = new ArrayList<>();
129          strings.add("Hello"); strings.add("World");
130       }
131    }
132
133    private static <T> String join(String separator, ArrayList<T> elements)
134    {
135       var result = new StringBuilder();
136       for (T e : elements)
137       {
138          if (result.length() > 0) result.append(separator);
139          result.append(e.toString());
140       }
141       return result.toString();
142    }
143
144    public static void main(String[] args) throws Exception
145    {
146       var formatter = new Formatter();
147       formatter.forType(new TypeLiteral<ArrayList<Integer>>(){},
148          lst -> join(" ", lst));
149       formatter.forType(new TypeLiteral<ArrayList<Character>>(){},
150          lst -> "\"" + join("", lst) + "\"");
151       System.out.println(formatter.formatFields(new Sample()));
152    }
153 }
```

---

java.lang.Class<T> 1.0

- TypeVariable[] getTypeParameters() 5

  gets the generic type variables if this type was declared as a generic type, or an array of length 0 otherwise.

- Type getGenericSuperclass() 5

  gets the generic type of the superclass that was declared for this type, or null if this type is Object or not a class type.

- Type[] getGenericInterfaces() 5

  gets the generic types of the interfaces that were declared for this type, in declaration order, or an array of length 0 if this type doesn't implement interfaces.

---

java.lang.reflect.Method 1.1

- TypeVariable[] getTypeParameters() 5

  gets the generic type variables if this method was declared as a generic method, or an array of length 0 otherwise.

- Type getGenericReturnType() 5

  gets the generic return type with which this method was declared.

- Type[] getGenericParameterTypes() 5

  gets the generic parameter types with which this method was declared. If the method has no parameters, an array of length 0 is returned.

---

java.lang.reflect.TypeVariable 5

- String getName()

  gets the name of this type variable.

- Type[] getBounds()

  gets the subclass bounds of this type variable, or an array of length 0 if the variable is unbounded.

---

java.lang.reflect.WildcardType 5

- Type[] getUpperBounds()

  gets the subclass (extends) bounds of this type variable, or an array of length 0 if the variable has no subclass bounds.

- Type[] getLowerBounds()

  gets the superclass (super) bounds of this type variable, or an array of length 0 if the variable has no superclass bounds.

---

java.lang.reflect.ParameterizedType 5

- Type getRawType()

  gets the raw type of this parameterized type.

- Type[] getActualTypeArguments()

  gets the type parameters with which this parameterized type was declared.

- Type getOwnerType()

  gets the outer class type if this is an inner type, or null if this is a top-level type.

---

java.lang.reflect.GenericArrayType 5

- Type getGenericComponentType()

  gets the generic component type with which this array type was declared.

You now know how to use generic classes and how to program your own generic classes and methods if the need arises. Just as importantly, you know how to decipher the generic type declarations that you may encounter in the API documentation and in error messages. For an exhaustive discussion of everything there is to know about Java generics, turn to Angelika Langer's excellent list of frequently (and not so frequently) asked questions at http://angelikalanger.com/GenericsFAQ/JavaGenericsFAQ.html.

In the next chapter, you will see how the Java collections framework puts generics to work.

CHAPTER 9

Collections

In this chapter

The data structures that you choose can make a big difference when you try to implement methods in a natural style or are concerned with performance. Do you need to search quickly through thousands (or even millions) of sorted items? Do you need to rapidly insert and remove elements in the middle of an ordered sequence? Do you need to establish associations between keys and values?

This chapter shows how the Java library can help you accomplish the traditional data structuring needed for serious programming. In college computer science programs, a course called *Data Structures* usually takes a semester to complete, and there are many, many books devoted to this important topic. Our coverage differs from that of a college course; we will skip the theory and just show you how to use the collection classes in the standard library.

9.1 The Java Collections Framework

The initial release of Java supplied only a small set of classes for the most useful data structures: Vector, Stack, Hashtable, BitSet, and the Enumeration interface that provides an abstract mechanism for visiting elements in an arbitrary container. That was certainly a wise choice—it takes time and skill to come up with a comprehensive collection class library.

With the advent of Java 1.2, the designers felt that the time had come to roll out a full-fledged set of data structures. They faced a number of conflicting design challenges. They wanted the library to be small and easy to learn. They did not want the complexity of the Standard Template Library (or STL) of C++, but they wanted the benefit of "generic algorithms" that STL pioneered. They wanted the legacy classes to fit into the new framework. As all designers of collections libraries do, they had to make some hard choices, and they came up with a number of idiosyncratic design decisions along the way. In this section, we will explore the basic design of the Java collections framework, show you how to put it to work, and explain the reasoning behind some of the more controversial features.

9.1.1 Separating Collection Interfaces and Implementation

As is common with modern data structure libraries, the Java collection library separates *interfaces* and *implementations*. Let us look at that separation with a familiar data structure, the *queue*.

A *queue interface* specifies that you can add elements at the tail end of the queue, remove them at the head, and find out how many elements are in the queue. You use a queue when you need to collect objects and retrieve them in a "first in, first out" fashion (see Figure 9.1).

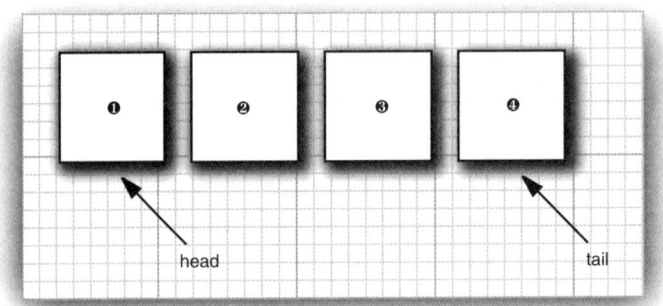

Figure 9.1 A queue

A minimal form of a queue interface might look like this:

```
public interface Queue<E> // a simplified form of the interface in the standard library
{
    void add(E element);
    E remove();
    int size();
}
```

The interface tells you nothing about how the queue is implemented. Of the two common implementations of a queue, one uses a "circular array" and one uses a linked list (see Figure 9.2).

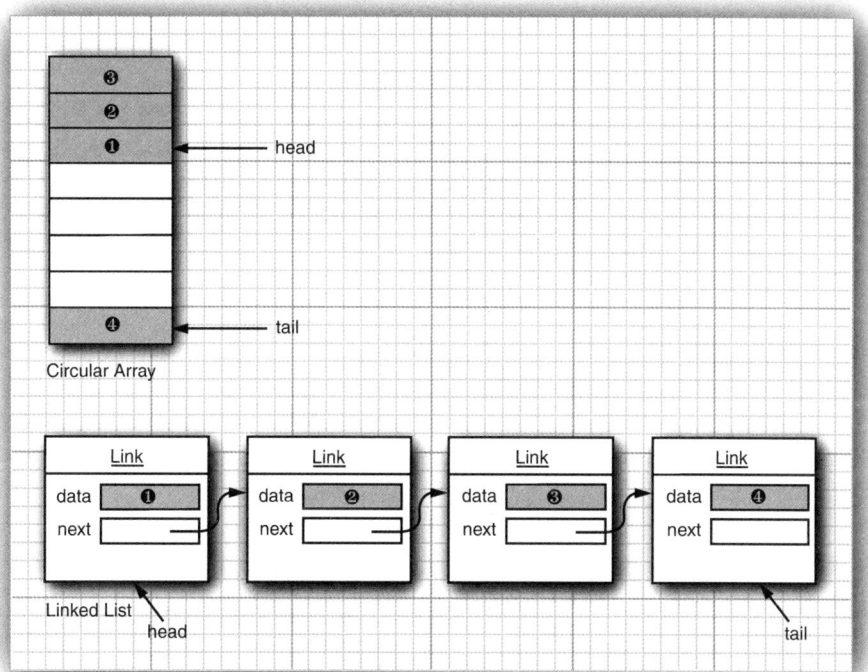

Figure 9.2 Queue implementations

Each implementation can be expressed by a class that implements the Queue interface.

```
public class CircularArrayQueue<E> implements Queue<E> // not an actual library class
{
    private int head;
    private int tail;
```

```
    CircularArrayQueue(int capacity) { . . . }
    public void add(E element) { . . . }
    public E remove() { . . . }
    public int size() { . . . }
    private E[] elements;
}

public class LinkedListQueue<E> implements Queue<E> // not an actual library class
{
    private Link head;
    private Link tail;

    LinkedListQueue() { . . . }
    public void add(E element) { . . . }
    public E remove() { . . . }
    public int size() { . . . }
}
```

 NOTE: The Java library doesn't actually have classes named `CircularArrayQueue` and `LinkedListQueue`. We use these classes as examples to explain the conceptual distinction between collection interfaces and implementations. If you need a circular array queue, use the `ArrayDeque` class. For a linked list queue, simply use the `LinkedList` class—it implements the `Queue` interface.

When you use a queue in your program, you don't need to know which implementation is actually used once the collection has been constructed. Therefore, it makes sense to use the concrete class *only* when you construct the collection object. Use the *interface type* to hold the collection reference.

```
Queue<Customer> expressLane = new CircularArrayQueue<>(100);
expressLane.add(new Customer("Harry"));
```

With this approach, if you change your mind, you can easily use a different implementation. You only need to change your program in one place—in the constructor call. If you decide that a `LinkedListQueue` is a better choice after all, your code becomes

```
Queue<Customer> expressLane = new LinkedListQueue<>();
expressLane.add(new Customer("Harry"));
```

Why would you choose one implementation over another? The interface says nothing about the efficiency of an implementation. A circular array is somewhat more efficient than a linked list, so it is generally preferable. However, as usual, there is a price to pay.

The circular array is a *bounded* collection—it has a finite capacity. If you don't have an upper limit on the number of objects that your program will collect, you may be better off with a linked list implementation after all.

When you study the API documentation, you will find another set of classes whose name begins with Abstract, such as AbstractQueue. These classes are intended for library implementors. In the (perhaps unlikely) event that you want to implement your own queue class, you will find it easier to extend AbstractQueue than to implement all the methods of the Queue interface.

9.1.2 The Collection Interface

The fundamental interface for collection classes in the Java library is the Collection interface. The interface has two fundamental methods:

```
public interface Collection<E>
{
   boolean add(E element);
   Iterator<E> iterator();
   . . .
}
```

There are several methods in addition to these two; we will discuss them later.

The add method adds an element to the collection. The add method returns true if adding the element actually changes the collection, and false if the collection is unchanged. For example, if you try to add an object to a set and the object is already present, the add request has no effect because sets reject duplicates.

The iterator method returns an object that implements the Iterator interface. You can use the iterator object to visit the elements in the collection one by one. We discuss iterators in the next section.

9.1.3 Iterators

The Iterator interface has four methods:

```
public interface Iterator<E>
{
   E next();
   boolean hasNext();
   void remove();
   default void forEachRemaining(Consumer<? super E> action);
}
```

By repeatedly calling the next method, you can visit the elements from the collection one by one. However, if you reach the end of the collection, the next method throws a NoSuchElementException. Therefore, you need to call the hasNext method before calling next. That method returns true if the iterator object still has more elements to visit. If you want to inspect all elements in a collection, request an iterator and then keep calling the next method while hasNext returns true. For example:

```
Collection<String> c = . . .;
Iterator<String> iter = c.iterator();
while (iter.hasNext())
{
    String element = iter.next();
    do something with element
}
```

You can write such a loop more concisely as the "for each" loop:

```
for (String element : c)
{
    do something with element
}
```

The compiler simply translates the "for each" loop into a loop with an iterator.

The "for each" loop works with any object that implements the Iterable interface, an interface with a single abstract method:

```
public interface Iterable<E>
{
    Iterator<E> iterator();
    . . .
}
```

The Collection interface extends the Iterable interface. Therefore, you can use the "for each" loop with any collection in the standard library.

Instead of writing a loop, you can call the forEachRemaining method with a lambda expression that consumes an element. The lambda expression is invoked with each element of the iterator, until there are none left.

```
iterator.forEachRemaining(element -> do something with element);
```

The order in which the elements are visited depends on the collection type. If you iterate over an ArrayList, the iterator starts at index 0 and increments the index in each step. However, if you visit the elements in a HashSet, you will get them in an essentially random order. You can be assured that you will encounter all elements of the collection during the course of the iteration, but you cannot make any assumptions about their ordering. This is usually not a problem because the ordering does not matter for computations such as computing totals or counting matches.

NOTE: Old-timers will notice that the next and hasNext methods of the Iterator interface serve the same purpose as the nextElement and hasMoreElements methods of an Enumeration. The designers of the Java collections library could have chosen to make use of the Enumeration interface. But they disliked the cumbersome method names and instead introduced a new interface with shorter method names.

There is an important conceptual difference between iterators in the Java collections library and iterators in other libraries. In traditional collections libraries, such as the Standard Template Library of C++, iterators are modeled after array indexes. Given such an iterator, you can look up the element that is stored at that position, much like you can look up an array element a[i] if you have an array index i. Independently of the lookup, you can advance the iterator to the next position. This is the same operation as advancing an array index by calling i++, without performing a lookup. However, the Java iterators do not work like that. The lookup and position change are tightly coupled. The only way to look up an element is to call next, and that lookup advances the position.

Instead, think of Java iterators as being *between elements*. When you call next, the iterator *jumps over* the next element, and it returns a reference to the element that it just passed (see Figure 9.3).

NOTE: Here is another useful analogy. You can think of Iterator.next as the equivalent of InputStream.read. Reading a byte from a stream automatically "consumes" the byte. The next call to read consumes and returns the next byte from the input. Similarly, repeated calls to next let you read all elements in a collection.

The remove method of the Iterator interface removes the element that was returned by the last call to next. In many situations, that makes sense—you need to see the element before you can decide that it is the one that should be removed. But if you want to remove an element in a particular position, you still need to skip past the element. For example, here is how you remove the first element in a collection of strings:

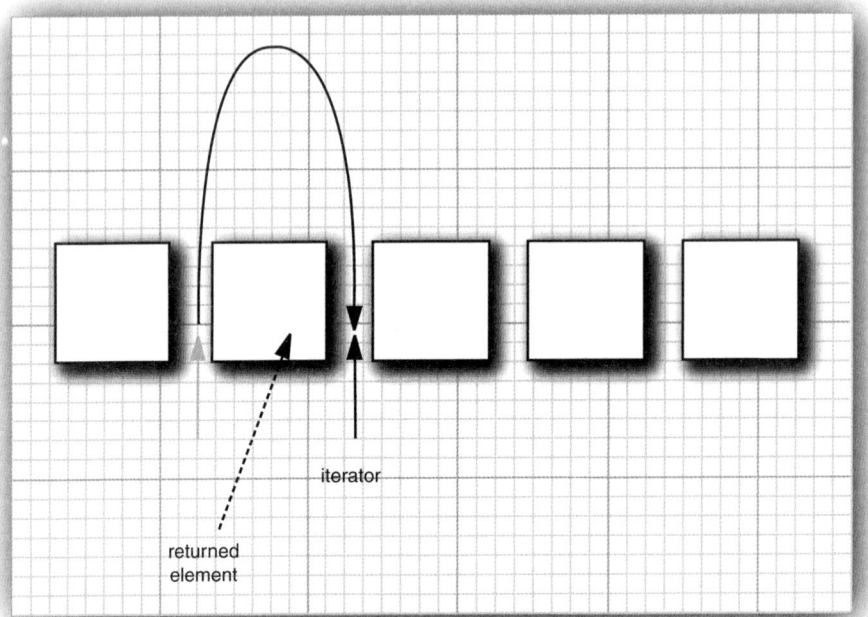

Figure 9.3 Advancing an iterator

```
Iterator<String> it = c.iterator();
it.next(); // skip over the first element
it.remove(); // now remove it
```

More importantly, there is a dependency between the calls to the next and remove methods. It is illegal to call remove if it wasn't preceded by a call to next. If you try, an IllegalStateException is thrown.

If you want to remove two adjacent elements, you cannot simply call

```
it.remove();
it.remove(); // ERROR
```

Instead, you must first call next to jump over the element to be removed.

```
it.remove();
it.next();
it.remove(); // OK
```

9.1.4 Generic Utility Methods

The Collection and Iterator interfaces are generic, which means you can write utility methods that operate on any kind of collection. For example, here is a generic method that tests whether an arbitrary collection contains a given element:

```
public static <E> boolean contains(Collection<E> c, Object obj)
{
   for (E element : c)
      if (element.equals(obj))
         return true;
   return false;
}
```

The designers of the Java library decided that some of these utility methods are so useful that the library should make them available. That way, library users don't have to keep reinventing the wheel. The contains method is one such method.

In fact, the Collection interface declares quite a few useful methods that all implementing classes must supply. Among them are

```
int size()
boolean isEmpty()
boolean contains(Object obj)
boolean containsAll(Collection<?> c)
boolean equals(Object other)
boolean addAll(Collection<? extends E> from)
boolean remove(Object obj)
boolean removeAll(Collection<?> c)
void clear()
boolean retainAll(Collection<?> c)
Object[] toArray()
<T> T[] toArray(T[] arrayToFill)
```

Many of these methods are self-explanatory; you will find full documentation in the API notes at the end of this section.

Of course, it is a bother if every class that implements the Collection interface has to supply so many routine methods. To make life easier for implementors, the library supplies a class AbstractCollection that leaves the fundamental methods size and iterator abstract but implements the routine methods in terms of them. For example:

```
public abstract class AbstractCollection<E>
    implements Collection<E>
{
   . . .
   public abstract Iterator<E> iterator();

   public boolean contains(Object obj)
   {
      for (E element : this) // calls iterator()
         if (element.equals(obj))
            return true;
      return false;
   }
   . . .
}
```

A concrete collection class can now extend the AbstractCollection class. It is up to the concrete collection class to supply an iterator method, but the contains method has been taken care of by the AbstractCollection superclass. However, if the subclass has a more efficient way of implementing contains, it is free to do so.

This approach is a bit outdated. It would be nicer if the methods were default methods of the Collection interface. This has not happened. However, several default methods have been added. Most of them deal with streams (which we will discuss in Volume II). In addition, there is a useful method

```
default boolean removeIf(Predicate<? super E> filter)
```

for removing elements that fulfill a condition.

---

java.util.Collection<E> 1.2

- Iterator<E> iterator()

  returns an iterator that can be used to visit the elements in the collection.
- int size()

  returns the number of elements currently stored in the collection.
- boolean isEmpty()

  returns true if this collection contains no elements.
- boolean contains(Object obj)

  returns true if this collection contains an object equal to obj.

(Continues)

---

java.util.Collection<E> 1.2 *(Continued)*

- boolean containsAll(Collection<?> other)

  returns true if this collection contains all elements in the other collection.

- boolean add(E element)

  adds an element to the collection. Returns true if the collection changed as a result of this call.

- boolean addAll(Collection<? extends E> other)

  adds all elements from the other collection to this collection. Returns true if the collection changed as a result of this call.

- boolean remove(Object obj)

  removes an object equal to obj from this collection. Returns true if a matching object was removed.

- boolean removeAll(Collection<?> other)

  removes from this collection all elements from the other collection. Returns true if the collection changed as a result of this call.

- default boolean removeIf(Predicate<? super E> filter) 8

  removes all elements for which filter returns true. Returns true if the collection changed as a result of this call.

- void clear()

  removes all elements from this collection.

- boolean retainAll(Collection<?> other)

  removes all elements from this collection that do not equal one of the elements in the other collection. Returns true if the collection changed as a result of this call.

- Object[] toArray()

  returns an array of the objects in the collection.

- <T> T[] toArray(T[] arrayToFill)

  returns an array of the objects in the collection. If arrayToFill has sufficient length, it is filled with the elements of this collection. If there is space, a null element is appended. Otherwise, a new array with the same component type as arrayToFill and the same length as the size of this collection is allocated and filled.

---

java.util.Iterator<E> 1.2

- boolean hasNext()

  returns true if there is another element to visit.

- E next()

  returns the next object to visit. Throws a NoSuchElementException if the end of the collection has been reached.

- void remove()

  removes the last visited object. This method must immediately follow an element visit. If the collection has been modified since the last element visit, this method throws an IllegalStateException.

- default void forEachRemaining(Consumer<? super E> action) 8

  visits elements and passes them to the given action until no elements remain or the action throws an exception.

9.2 Interfaces in the Collections Framework

The Java collections framework defines a number of interfaces for different types of collections, shown in Figure 9.4.

There are two fundamental interfaces for collections: Collection and Map. As you already saw, you insert elements into a collection with a method

```
boolean add(E element)
```

However, maps hold key/value pairs, and you use the put method to insert them:

```
V put(K key, V value)
```

To read elements from a collection, visit them with an iterator. However, you can read values from a map with the get method:

```
V get(K key)
```

A List is an *ordered collection*. Elements are added into a particular position in the container. An element can be accessed in two ways: by an iterator or by an integer index. The latter is called *random access* because elements can be visited in any order. In contrast, when using an iterator, one must visit them sequentially.

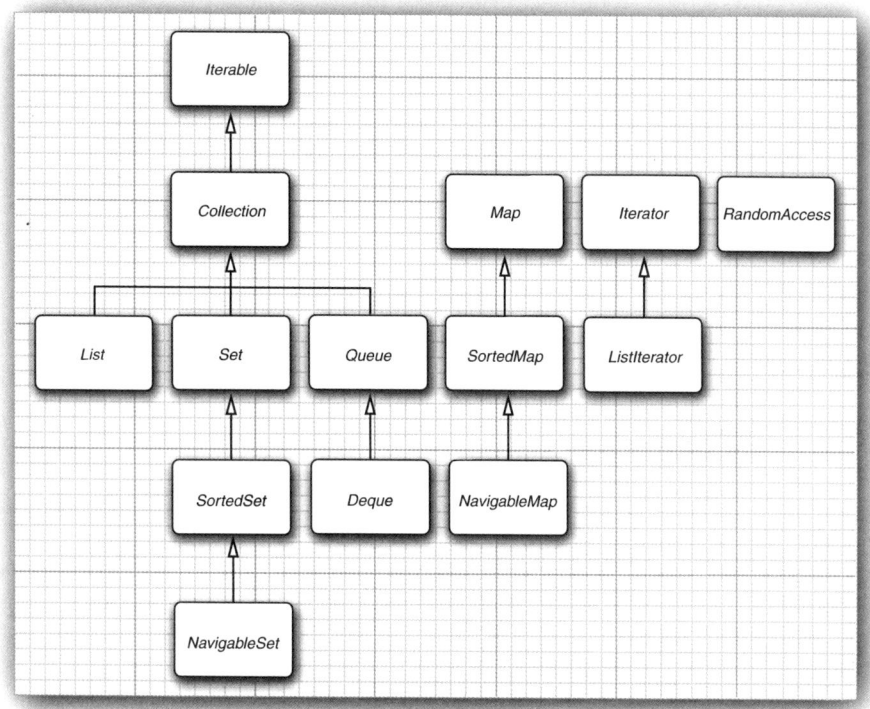

Figure 9.4 The interfaces of the collections framework

The List interface defines several methods for random access:

```
void add(int index, E element)
void remove(int index)
E get(int index)
E set(int index, E element)
```

The ListIterator interface is a subinterface of Iterator. It defines a method for adding an element before the iterator position:

```
void add(E element)
```

Frankly, this aspect of the collections framework is poorly designed. In practice, there are two kinds of ordered collections, with very different performance tradeoffs. An ordered collection that is backed by an array has fast random access, and it makes sense to use the List methods with an integer index. In contrast, a linked list, while also ordered, has slow random access, and it is best traversed with an iterator. It would have been an easy matter to provide two interfaces.

 NOTE: To avoid carrying out random access operations for linked lists, Java 1.4 introduced a tagging interface, RandomAccess. That interface has no methods, but you can use it to test whether a particular collection supports efficient random access:

```
if (c instanceof RandomAccess)
{
    use random access algorithm
}
else
{
    use sequential access algorithm
}
```

The Set interface is identical to the Collection interface, but the behavior of the methods is more tightly defined. The add method of a set should reject duplicates. The equals method of a set should be defined so that two sets are identical if they have the same elements, but not necessarily in the same order. The hashCode method should be defined so that two sets with the same elements yield the same hash code.

Why make a separate interface if the method signatures are the same? Conceptually, not all collections are sets. Making a Set interface enables programmers to write methods that accept only sets.

The SortedSet and SortedMap interfaces expose the comparator object used for sorting, and they define methods to obtain views of subsets of the collections. We discuss these in Section 9.5, "Views and Wrappers," on p. 532.

Finally, Java 6 introduced interfaces NavigableSet and NavigableMap that contain additional methods for searching and traversal in sorted sets and maps. (Ideally, these methods should have simply been included in the SortedSet and SortedMap interface.) The TreeSet and TreeMap classes implement these interfaces.

9.3 Concrete Collections

Table 9.1 shows the collections in the Java library and briefly describes the purpose of each collection class. (For simplicity, we omit the thread-safe collections that will be discussed in Chapter 12.) All classes in Table 9.1 implement the Collection interface, with the exception of the classes with names ending in Map. Those classes implement the Map interface instead. We will discuss maps in Section 9.4, "Maps," on p. 519.

Figure 9.5 shows the relationships between these classes.

Table 9.1 Concrete Collections in the Java Library

Collection Type	Description	See Page
ArrayList	An indexed sequence that grows and shrinks dynamically	507
LinkedList	An ordered sequence that allows efficient insertion and removal at any location	496
ArrayDeque	A double-ended queue that is implemented as a circular array	516
HashSet	An unordered collection that rejects duplicates	507
TreeSet	A sorted set	511
EnumSet	A set of enumerated type values	529
LinkedHashSet	A set that remembers the order in which elements were inserted	527
PriorityQueue	A collection that allows efficient removal of the smallest element	518
HashMap	A data structure that stores key/value associations	526
TreeMap	A map in which the keys are sorted	519
EnumMap	A map in which the keys belong to an enumerated type	529
LinkedHashMap	A map that remembers the order in which entries were added	527
WeakHashMap	A map with values that can be reclaimed by the garbage collector if they are not used elsewhere	526
IdentityHashMap	A map with keys that are compared by ==, not equals	530

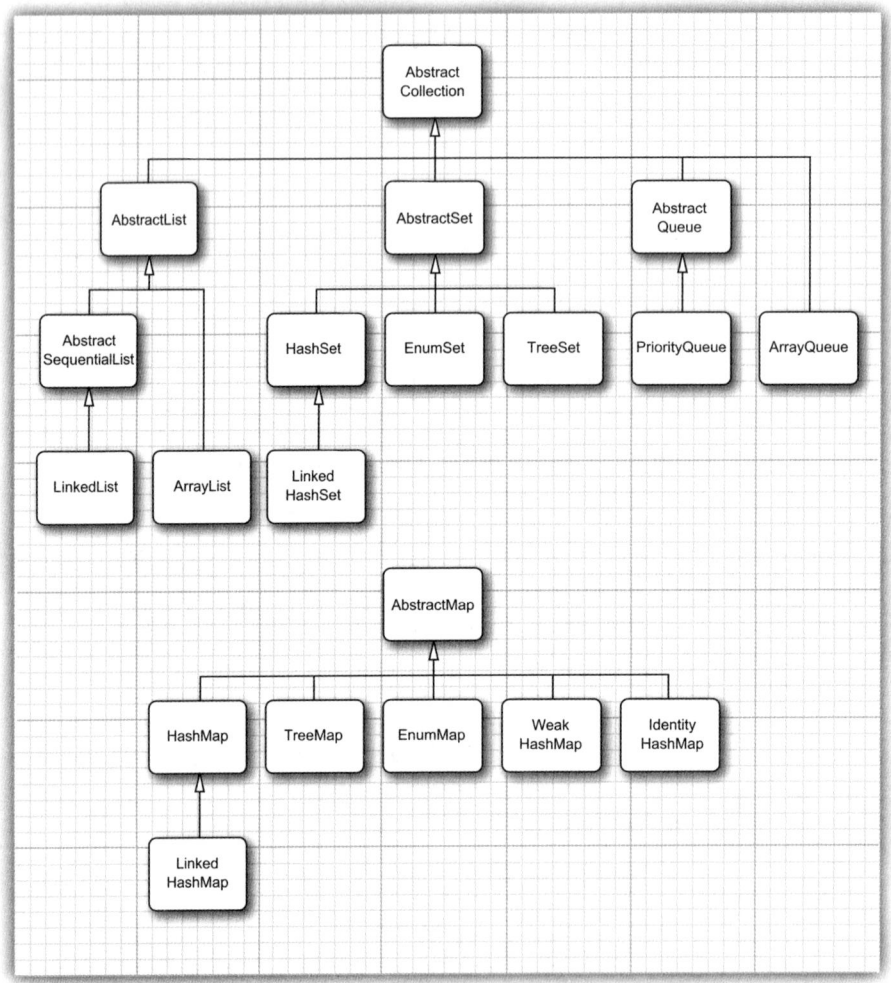

Figure 9.5 Classes in the collections framework

9.3.1 Linked Lists

We already used arrays and their dynamic cousin, the ArrayList class, for many examples in this book. However, arrays and array lists suffer from a major drawback. Removing an element from the middle of an array is expensive since all array elements beyond the removed one must be moved toward the beginning of the array (see Figure 9.6). The same is true for inserting elements in the middle.

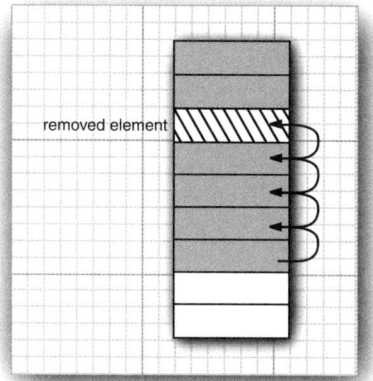

Figure 9.6 Removing an element from an array

Another well-known data structure, the *linked list*, solves this problem. Where an array stores object references in consecutive memory locations, a linked list stores each object in a separate *link*. Each link also stores a reference to the next link in the sequence. In the Java programming language, all linked lists are actually *doubly linked*; that is, each link also stores a reference to its predecessor (see Figure 9.7).

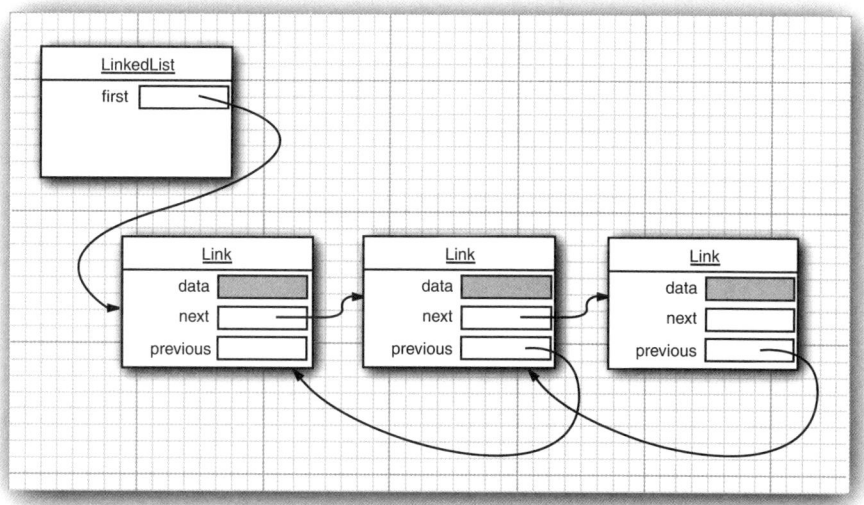

Figure 9.7 A doubly linked list

Removing an element from the middle of a linked list is an inexpensive operation—only the links around the element to be removed need to be updated (see Figure 9.8).

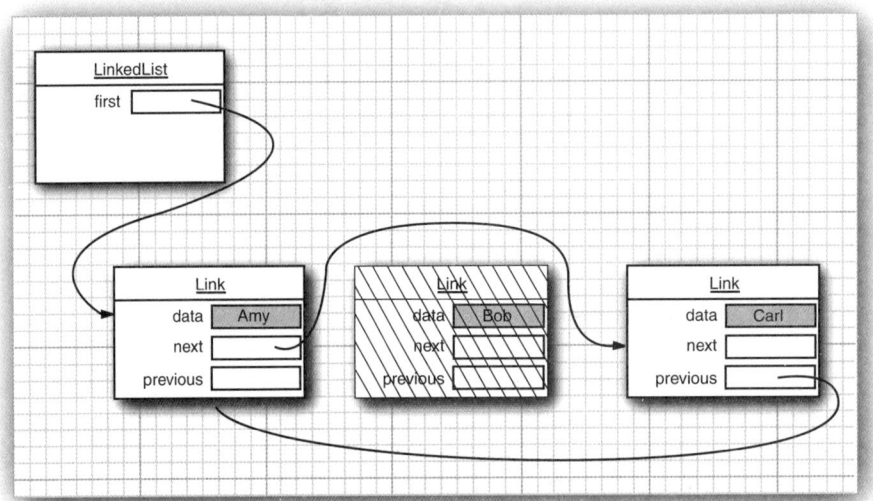

Figure 9.8 Removing an element from a linked list

Perhaps you once took a data structures course in which you learned how to implement linked lists. You may have bad memories of tangling up the links when removing or adding elements in the linked list. If so, you will be pleased to learn that the Java collections library supplies a class LinkedList ready for you to use.

The following code example adds three elements and then removes the second one:

```
var staff = new LinkedList<String>();
staff.add("Amy");
staff.add("Bob");
staff.add("Carl");
Iterator<String> iter = staff.iterator();
String first = iter.next(); // visit first element
String second = iter.next(); // visit second element
iter.remove(); // remove last visited element
```

There is, however, an important difference between linked lists and generic collections. A linked list is an *ordered collection* in which the position of the objects matters. The LinkedList.add method adds the object to the end of the list.

But you will often want to add objects somewhere in the middle of a list. This position-dependent add method is the responsibility of an iterator, since iterators describe positions in collections. Using iterators to add elements makes sense only for collections that have a natural ordering. For example, the *set* data type that we discuss in the next section does not impose any ordering on its elements. Therefore, there is no add method in the Iterator interface. Instead, the collections library supplies a subinterface ListIterator that contains an add method:

```
interface ListIterator<E> extends Iterator<E>
{
   void add(E element);
   . . .
}
```

Unlike Collection.add, this method does not return a boolean—it is assumed that the add operation always modifies the list.

In addition, the ListIterator interface has two methods that you can use for traversing a list backwards.

```
E previous()
boolean hasPrevious()
```

Like the next method, the previous method returns the object that it skipped over.

The listIterator method of the LinkedList class returns an iterator object that implements the ListIterator interface.

```
ListIterator<String> iter = staff.listIterator();
```

The add method adds the new element *before* the iterator position. For example, the following code skips past the first element in the linked list and adds "Juliet" before the second element (see Figure 9.9):

```
var staff = new LinkedList<String>();
staff.add("Amy");
staff.add("Bob");
staff.add("Carl");
ListIterator<String> iter = staff.listIterator();
iter.next(); // skip past first element
iter.add("Juliet");
```

If you call the add method multiple times, the elements are simply added in the order in which you supplied them. They are all added in turn before the current iterator position.

When you use the add operation with an iterator that was freshly returned from the listIterator method and that points to the beginning of the linked list, the newly added element becomes the new head of the list. When the

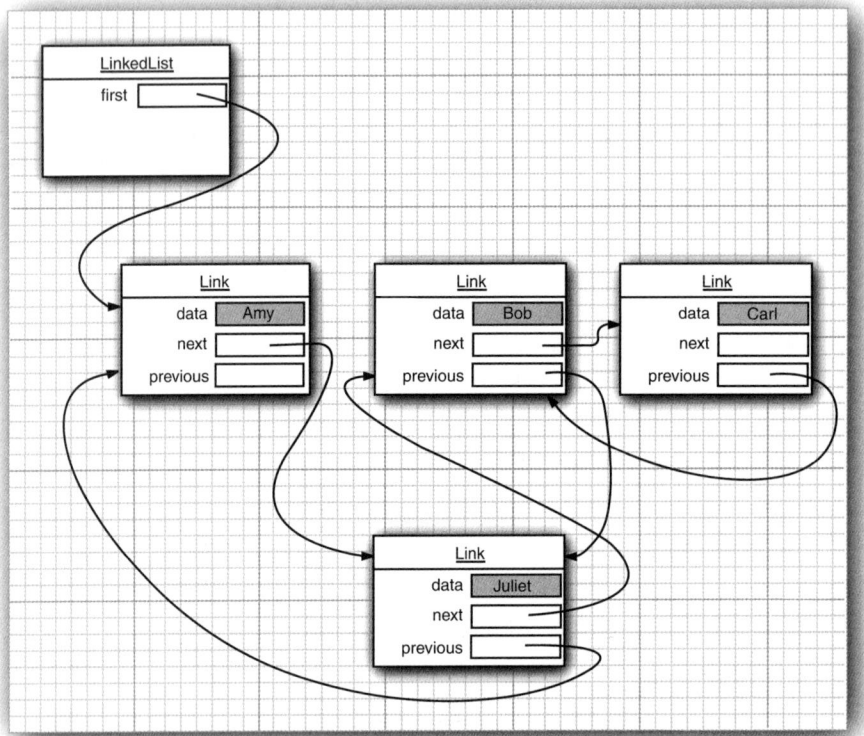

Figure 9.9 Adding an element to a linked list

iterator has passed the last element of the list (that is, when hasNext returns false), the added element becomes the new tail of the list. If the linked list has *n* elements, there are *n* + 1 spots for adding a new element. These spots correspond to the *n* + 1 possible positions of the iterator. For example, if a linked list contains three elements, A, B, and C, there are four possible positions (marked as |) for inserting a new element:

```
|ABC
A|BC
AB|C
ABC|
```

NOTE: Be careful with the "cursor" analogy. The remove operation does not work exactly like the Backspace key. Immediately after a call to next, the remove method indeed removes the element to the left of the iterator, just like the Backspace key would. However, if you have just called previous, the element to the right will be removed. And you can't call remove twice in a row.

Unlike the add method, which depends only on the iterator position, the remove method depends on the iterator state.

Finally, a set method replaces the last element, returned by a call to next or previous, with a new element. For example, the following code replaces the first element of a list with a new value:

```
ListIterator<String> iter = list.listIterator();
String oldValue = iter.next(); // returns first element
iter.set(newValue); // sets first element to newValue
```

As you might imagine, if an iterator traverses a collection while another iterator is modifying it, confusing situations can occur. For example, suppose an iterator points before an element that another iterator has just removed. The iterator is now invalid and should no longer be used. The linked list iterators have been designed to detect such modifications. If an iterator finds that its collection has been modified by another iterator or by a method of the collection itself, it throws a ConcurrentModificationException. For example, consider the following code:

```
List<String> list = . . .;
ListIterator<String> iter1 = list.listIterator();
ListIterator<String> iter2 = list.listIterator();
iter1.next();
iter1.remove();
iter2.next(); // throws ConcurrentModificationException
```

The call to iter2.next throws a ConcurrentModificationException since iter2 detects that the list was modified externally.

To avoid concurrent modification exceptions, follow this simple rule: You can attach as many iterators to a collection as you like, provided that all of them are only readers. Alternatively, you can attach a single iterator that can both read and write.

Concurrent modification detection is done in a simple way. The collection keeps track of the number of mutating operations (such as adding and removing elements). Each iterator keeps a separate count of the number of mutating operations that *it* was responsible for. At the beginning of each iterator method, the iterator simply checks whether its own mutation count equals that of the collection. If not, it throws a ConcurrentModificationException.

 NOTE: There is, however, a curious exception to the detection of concurrent modifications. The linked list only keeps track of *structural* modifications to the list, such as adding and removing links. The set method does *not* count as a structural modification. You can attach multiple iterators to a linked list, all of which call set to change the contents of existing links. This capability is required for a number of algorithms in the Collections class that we discuss later in this chapter.

Now you have seen the fundamental methods of the LinkedList class. Use a ListIterator to traverse the elements of the linked list in either direction and to add and remove elements.

As you saw in Section 9.2, "Interfaces in the Collections Framework," on p. 492, many other useful methods for operating on linked lists are declared in the Collection interface. These are, for the most part, implemented in the AbstractCollection superclass of the LinkedList class. For example, the toString method invokes toString on all elements and produces one long string of the format [A, B, C]. This is handy for debugging. Use the contains method to check whether an element is present in a linked list. For example, the call staff.contains("Harry") returns true if the linked list already contains a string equal to the string "Harry".

The library also supplies a number of methods that are, from a theoretical perspective, somewhat dubious. Linked lists do not support fast random access. If you want to see the *n*th element of a linked list, you have to start at the beginning and skip past the first $n - 1$ elements. There is no shortcut. For that reason, programmers don't usually use linked lists in situations where elements need to be accessed by an integer index.

Nevertheless, the LinkedList class supplies a get method that lets you access a particular element:

```
LinkedList<String> list = . . .;
String obj = list.get(n);
```

Of course, this method is not very efficient. If you find yourself using it, you are probably using a wrong data structure for your problem.

You should *never* use this illusory random access method to step through a linked list. The code

```
for (int i = 0; i < list.size(); i++)
    do something with list.get(i);
```

is staggeringly inefficient. Each time you look up another element, the search starts again from the beginning of the list. The LinkedList object makes no effort to cache the position information.

 NOTE: The get method has one slight optimization: If the index is at least size() / 2, the search for the element starts at the end of the list.

The list iterator interface also has a method to tell you the index of the current position. In fact, since Java iterators conceptually point between elements, it has two of them: The nextIndex method returns the integer index of the element that would be returned by the next call to next; the previousIndex method returns the index of the element that would be returned by the next call to previous. Of course, that is simply one less than nextIndex. These methods are efficient—an iterator keeps a count of its current position. Finally, if you have an integer index n, then list.listIterator(n) returns an iterator that points just before the element with index n. That is, calling next yields the same element as list.get(n); obtaining that iterator is inefficient.

If you have a linked list with only a handful of elements, you don't have to be overly paranoid about the cost of the get and set methods. But then, why use a linked list in the first place? The only reason to use a linked list is to minimize the cost of insertion and removal in the middle of the list. If you have only a few elements, you can just use an ArrayList.

We recommend that you simply stay away from all methods that use an integer index to denote a position in a linked list. If you want random access into a collection, use an array or ArrayList, not a linked list.

The program in Listing 9.1 puts linked lists to work. It simply creates two lists, merges them, then removes every second element from the second list, and finally tests the removeAll method. We recommend that you trace the program flow and pay special attention to the iterators. You may find it helpful to draw diagrams of the iterator positions, like this:

```
|ACE  |BDFG
A|CE  |BDFG
AB|CE B|DFG
 . . .
```

Note that the call

```
System.out.println(a);
```

prints all elements in the linked list a by invoking the toString method in AbstractCollection.

Listing 9.1 linkedList/LinkedListTest.java

```
1  package linkedList;
2
3  import java.util.*;
4
5  /**
6   * This program demonstrates operations on linked lists.
7   * @version 1.12 2018-04-10
8   * @author Cay Horstmann
9   */
10  public class LinkedListTest
11  {
12      public static void main(String[] args)
13      {
14          var a = new LinkedList<String>();
15          a.add("Amy");
16          a.add("Carl");
17          a.add("Erica");
18
19          var b = new LinkedList<String>();
20          b.add("Bob");
21          b.add("Doug");
22          b.add("Frances");
23          b.add("Gloria");
24
25          // merge the words from b into a
26
27          ListIterator<String> aIter = a.listIterator();
28          Iterator<String> bIter = b.iterator();
29
30          while (bIter.hasNext())
31          {
32              if (aIter.hasNext()) aIter.next();
33              aIter.add(bIter.next());
34          }
35
36          System.out.println(a);
37
38          // remove every second word from b
39
40          bIter = b.iterator();
41          while (bIter.hasNext())
42          {
43              bIter.next(); // skip one element
44              if (bIter.hasNext())
45              {
```

```
46              bIter.next(); // skip next element
47              bIter.remove(); // remove that element
48          }
49      }
50
51      System.out.println(b);
52
53      // bulk operation: remove all words in b from a
54
55      a.removeAll(b);
56
57      System.out.println(a);
58  }
59 }
```

---

java.util.List<E> 1.2

- ListIterator<E> listIterator()

  returns a list iterator for visiting the elements of the list.

- ListIterator<E> listIterator(int index)

  returns a list iterator for visiting the elements of the list whose first call to next will return the element with the given index.

- void add(int i, E element)

  adds an element at the specified position.

- void addAll(int i, Collection<? extends E> elements)

  adds all elements from a collection to the specified position.

- E remove(int i)

  removes and returns the element at the specified position.

- E get(int i)

  gets the element at the specified position.

- E set(int i, E element)

  replaces the element at the specified position with a new element and returns the old element.

- int indexOf(Object element)

  returns the position of the first occurrence of an element equal to the specified element, or -1 if no matching element is found.

- int lastIndexOf(Object element)

  returns the position of the last occurrence of an element equal to the specified element, or -1 if no matching element is found.

---

java.util.ListIterator<E> 1.2

- void add(E newElement)

  adds an element before the current position.

- void set(E newElement)

  replaces the last element visited by next or previous with a new element. Throws an IllegalStateException if the list structure was modified since the last call to next or previous.

- boolean hasPrevious()

  returns true if there is another element to visit when iterating backwards through the list.

- E previous()

  returns the previous object. Throws a NoSuchElementException if the beginning of the list has been reached.

- int nextIndex()

  returns the index of the element that would be returned by the next call to next.

- int previousIndex()

  returns the index of the element that would be returned by the next call to previous.

---

java.util.LinkedList<E> 1.2

- LinkedList()

  constructs an empty linked list.

- LinkedList(Collection<? extends E> elements)

  constructs a linked list and adds all elements from a collection.

- void addFirst(E element)
- void addLast(E element)

  adds an element to the beginning or the end of the list.

- E getFirst()
- E getLast()

  returns the element at the beginning or the end of the list.

- E removeFirst()
- E removeLast()

  removes and returns the element at the beginning or the end of the list.

9.3.2 Array Lists

In the preceding section, you saw the List interface and the LinkedList class that implements it. The List interface describes an ordered collection in which the position of elements matters. There are two protocols for visiting the elements: through an iterator and by random access with methods get and set. The latter is not appropriate for linked lists, but of course get and set make a lot of sense for arrays. The collections library supplies the familiar ArrayList class that also implements the List interface. An ArrayList encapsulates a dynamically reallocated array of objects.

 NOTE: If you are a veteran Java programmer, you may have used the Vector class whenever you need a dynamic array. Why use an ArrayList instead of a Vector? For one simple reason: All methods of the Vector class are *synchronized*. It is safe to access a Vector object from two threads. But if you access a vector from only a single thread—by far the more common case—your code wastes quite a bit of time with synchronization. In contrast, the ArrayList methods are not synchronized. We recommend that you use an ArrayList instead of a Vector whenever you don't need synchronization.

9.3.3 Hash Sets

Linked lists and arrays let you specify the order in which you want to arrange the elements. However, if you are looking for a particular element and don't remember its position, you need to visit all elements until you find a match. That can be time consuming if the collection contains many elements. If you don't care about the ordering of the elements, there are data structures that let you find elements much faster. The drawback is that those data structures give you no control over the order in which the elements appear. These data structures organize the elements in an order that is convenient for their own purposes.

A well-known data structure for finding objects quickly is the *hash table*. A hash table computes an integer, called the *hash code*, for each object. A hash code is somehow derived from the instance fields of an object, preferably in such a way that objects with different data yield different codes. Table 9.2 lists a few examples of hash codes that result from the hashCode method of the String class.

If you define your own classes, you are responsible for implementing your own hashCode method—see Chapter 5 for more information. Your implementation needs to be compatible with the equals method: If a.equals(b), then a and b must have the same hash code.

Table 9.2 Hash Codes Resulting from the hashCode Method

String	Hash Code
"Lee"	76268
"lee"	107020
"eel"	100300

What's important for now is that hash codes can be computed quickly and that the computation depends only on the state of the object that needs to be hashed, not on the other objects in the hash table.

In Java, hash tables are implemented as arrays of linked lists. Each list is called a *bucket* (see Figure 9.10). To find the place of an object in the table, compute its hash code and reduce it modulo the total number of buckets. The resulting number is the index of the bucket that holds the element. For example, if an object has hash code 76268 and there are 128 buckets, then the object is placed in bucket 108 (because the remainder 76268 % 128 is 108). Perhaps you are lucky and there is no other element in that bucket. Then, you simply insert the element into that bucket. Of course, sometimes you will hit a bucket that is already filled. This is called a *hash collision*. Then, compare the new object with all objects in that bucket to see if it is already present. If the hash codes are reasonably randomly distributed and the number of buckets is large enough, only a few comparisons should be necessary.

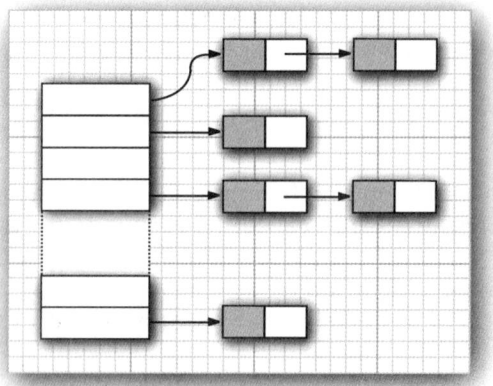

Figure 9.10 A hash table

> **NOTE:** As of Java 8, the buckets change from linked lists into balanced binary trees when they get full. This improves performance if a hash function was poorly chosen and yields many collisions, or if malicious code tries to flood a hash table with many values that have identical hash codes.

If you want more control over the performance of the hash table, you can specify the initial bucket count. The bucket count gives the number of buckets used to collect objects with identical hash values. If too many elements are inserted into a hash table, the number of collisions increases and retrieval performance suffers.

If you know how many elements, approximately, will eventually be in the table, you can set the bucket count. Typically, you should set it to somewhere between 75% and 150% of the expected element count. Some researchers believe that it is a good idea to make the bucket count a prime number to prevent a clustering of keys. The evidence for this isn't conclusive, however. The standard library uses bucket counts that are powers of 2, with a default of 16. (Any value you supply for the table size is automatically rounded to the next power of 2.)

Of course, you do not always know how many elements you need to store, or your initial guess may be too low. If the hash table gets too full, it needs to be *rehashed*. To rehash the table, a table with more buckets is created, all elements are inserted into the new table, and the original table is discarded. The *load factor* determines when a hash table is rehashed. For example, if the load factor is 0.75 (which is the default) and the table is more than 75% full, it is automatically rehashed with twice as many buckets. For most applications, it is reasonable to leave the load factor at 0.75.

Hash tables can be used to implement several important data structures. The simplest among them is the *set* type. A set is a collection of elements without duplicates. The add method of a set first tries to find the object to be added, and adds it only if it is not yet present.

The Java collections library supplies a HashSet class that implements a set based on a hash table. You add elements with the add method. The contains method is redefined to make a fast lookup to see if an element is already present in the set. It checks only the elements in one bucket and not all elements in the collection.

The hash set iterator visits all buckets in turn. Since hashing scatters the elements around in the table, they are visited in a seemingly random order. You would only use a HashSet if you don't care about the ordering of the elements in the collection.

The sample program at the end of this section (Listing 9.2) reads words from System.in, adds them to a set, and finally prints out the first twenty words in the set. For example, you can feed the program the text from *Alice in Wonderland* (which you can obtain from www.gutenberg.org) by launching it from a command shell as

```
java SetTest < alice30.txt
```

The program reads all words from the input and adds them to the hash set. It then iterates through the unique words in the set and finally prints out a count. (*Alice in Wonderland* has 5,909 unique words, including the copyright notice at the beginning.) The words appear in random order.

 CAUTION: Be careful when you mutate set elements. If the hash code of an element were to change, the element would no longer be in the correct position in the data structure.

Listing 9.2 set/SetTest.java

```java
1  package set;
2
3  import java.util.*;
4
5  /**
6   * This program uses a set to print all unique words in System.in.
7   * @version 1.12 2015-06-21
8   * @author Cay Horstmann
9   */
10 public class SetTest
11 {
12    public static void main(String[] args)
13    {
14       var words = new HashSet<String>();
15       long totalTime = 0;
16
17       try (var in = new Scanner(System.in))
18       {
19          while (in.hasNext())
20          {
21             String word = in.next();
22             long callTime = System.currentTimeMillis();
23             words.add(word);
24             callTime = System.currentTimeMillis() - callTime;
25             totalTime += callTime;
26          }
27       }
28
```

```
29    Iterator<String> iter = words.iterator();
30    for (int i = 1; i <= 20 && iter.hasNext(); i++)
31        System.out.println(iter.next());
32    System.out.println(". . .");
33    System.out.println(words.size() + " distinct words. " + totalTime + " milliseconds.");
34    }
35 }
```

---

java.util.HashSet<E> 1.2

- HashSet()

  constructs an empty hash set.

- HashSet(Collection<? extends E> elements)

  constructs a hash set and adds all elements from a collection.

- HashSet(int initialCapacity)

  constructs an empty hash set with the specified capacity (number of buckets).

- HashSet(int initialCapacity, float loadFactor)

  constructs an empty hash set with the specified capacity and load factor (a number between 0.0 and 1.0 that determines at what percentage of fullness the hash table will be rehashed into a larger one).

---

java.lang.Object 1.0

- int hashCode()

  returns a hash code for this object. A hash code can be any integer, positive or negative. The definitions of equals and hashCode must be compatible: If x.equals(y) is true, then x.hashCode() must be the same value as y.hashCode().

9.3.4 Tree Sets

The TreeSet class is similar to the hash set, with one added improvement. A tree set is a *sorted collection*. You insert elements into the collection in any order. When you iterate through the collection, the values are automatically presented in sorted order. For example, suppose you insert three strings and then visit all elements that you added.

```
var sorter = new TreeSet<String>();
sorter.add("Bob");
sorter.add("Amy");
sorter.add("Carl");
for (String s : sorter) System.out.println(s);
```

Then, the values are printed in sorted order: Amy Bob Carl. As the name of the class suggests, the sorting is accomplished by a tree data structure. (The current implementation uses a *red-black tree*. For a detailed description of red-black trees see, for example, *Introduction to Algorithms* by Thomas Cormen, Charles Leiserson, Ronald Rivest, and Clifford Stein, The MIT Press, 2009.) Every time an element is added to a tree, it is placed into its proper sorting position. Therefore, the iterator always visits the elements in sorted order.

Adding an element to a tree is slower than adding it to a hash table—see Table 9.3 for a comparison. But it is still much faster than checking for duplicates in an array or linked list. If the tree contains n elements, then an average of $\log_2 n$ comparisons are required to find the correct position for the new element. For example, if the tree already contains 1,000 elements, adding a new element requires about 10 comparisons.

 NOTE: In order to use a tree set, you must be able to compare the elements. The elements must implement the Comparable interface, or you must supply a Comparator when constructing the set. (The Comparable and Comparator interfaces were introduced in Chapter 6.)

Table 9.3 Adding Elements into Hash and Tree Sets

Document	Total Number of Words	Number of Distinct Words	HashSet	TreeSet
Alice in Wonderland	28195	5909	5 sec	7 sec
The Count of Monte Cristo	466300	37545	75 sec	98 sec

If you look back at Table 9.3, you may well wonder if you should always use a tree set instead of a hash set. After all, adding elements does not seem to take much longer, and the elements are automatically sorted. The answer depends on the data that you are collecting. If you don't need the data sorted, there is no reason to pay for the sorting overhead. More important, with some data it is much more difficult to come up with a sort order than a hash function. A hash function only needs to do a reasonably good job of scrambling the objects, whereas a comparison function must tell objects apart with complete precision.

To make this distinction more concrete, consider the task of collecting a set of rectangles. If you use a TreeSet, you need to supply a Comparator<Rectangle>. How do you compare two rectangles? By area? That doesn't work. You can have two different rectangles with different coordinates but the same area. The sort order for a tree must be a *total ordering*. Any two elements must be comparable, and the comparison can only be zero if the elements are equal. There is such a sort order for rectangles (the lexicographic ordering on its coordinates), but it is unnatural and cumbersome to compute. In contrast, a hash function is already defined for the Rectangle class. It simply hashes the coordinates.

 NOTE: As of Java 6, the TreeSet class implements the NavigableSet interface. That interface adds several convenient methods for locating elements and for backward traversal. See the API notes for details.

The program in Listing 9.3 builds two tree sets of Item objects. The first one is sorted by part number, the default sort order of Item objects. The second set is sorted by description, using a custom comparator.

Listing 9.3 treeSet/TreeSetTest.java

```
1  package treeSet;
2
3  import java.util.*;
4
5  /**
6   * This program sorts a set of Item objects by comparing their descriptions.
7   * @version 1.13 2018-04-10
8   * @author Cay Horstmann
9   */
10 public class TreeSetTest
11 {
12    public static void main(String[] args)
13    {
14       var parts = new TreeSet<Item>();
15       parts.add(new Item("Toaster", 1234));
16       parts.add(new Item("Widget", 4562));
17       parts.add(new Item("Modem", 9912));
18       System.out.println(parts);
19
20       var sortByDescription = new TreeSet<Item>(Comparator.comparing(Item::getDescription));
21
22       sortByDescription.addAll(parts);
23       System.out.println(sortByDescription);
24    }
25 }
```

Listing 9.4 treeSet/Item.java

```java
1  package treeSet;
2
3  import java.util.*;
4
5  /**
6   * An item with a description and a part number.
7   */
8  public class Item implements Comparable<Item>
9  {
10     private String description;
11     private int partNumber;
12
13     /**
14      * Constructs an item.
15      * @param aDescription the item's description
16      * @param aPartNumber the item's part number
17      */
18     public Item(String aDescription, int aPartNumber)
19     {
20        description = aDescription;
21        partNumber = aPartNumber;
22     }
23
24     /**
25      * Gets the description of this item.
26      * @return the description
27      */
28     public String getDescription()
29     {
30        return description;
31     }
32
33     public String toString()
34     {
35        return "[description=" + description + ", partNumber=" + partNumber + "]";
36     }
37
38     public boolean equals(Object otherObject)
39     {
40        if (this == otherObject) return true;
41        if (otherObject == null) return false;
```

```
42        if (getClass() != otherObject.getClass()) return false;
43        var other = (Item) otherObject;
44        return Objects.equals(description, other.description) && partNumber == other.partNumber;
45     }
46
47     public int hashCode()
48     {
49        return Objects.hash(description, partNumber);
50     }
51
52     public int compareTo(Item other)
53     {
54        int diff = Integer.compare(partNumber, other.partNumber);
55        return diff != 0 ? diff : description.compareTo(other.description);
56     }
57  }
```

---

java.util.TreeSet<E> 1.2

- TreeSet()
- TreeSet(Comparator<? super E> comparator)

  constructs an empty tree set.

- TreeSet(Collection<? extends E> elements)
- TreeSet(SortedSet<E> s)

  constructs a tree set and adds all elements from a collection or sorted set (in the latter case, using the same ordering).

---

java.util.SortedSet<E> 1.2

- Comparator<? super E> comparator()

  returns the comparator used for sorting the elements, or null if the elements are compared with the compareTo method of the Comparable interface.

- E first()
- E last()

  returns the smallest or largest element in the sorted set.

---

java.util.NavigableSet<E> 6

- E higher(E value)
- E lower(E value)

  returns the least element > value or the largest element < value, or null if there is no such element.

- E ceiling(E value)
- E floor(E value)

  returns the least element ≥ value or the largest element ≤ value, or null if there is no such element.

- E pollFirst()
- E pollLast()

  removes and returns the smallest or largest element in this set, or null if the set is empty.

- Iterator<E> descendingIterator()

  returns an iterator that traverses this set in descending direction.

9.3.5 Queues and Deques

As we already discussed, a queue lets you efficiently add elements at the tail and remove elements from the head. A double-ended queue, or *deque*, lets you efficiently add or remove elements at the head and tail. Adding elements in the middle is not supported. Java 6 introduced a Deque interface. It is implemented by the ArrayDeque and LinkedList classes, both of which provide deques whose size grows as needed. In Chapter 12, you will see bounded queues and deques.

---

java.util.Queue<E> 5

- boolean add(E element)
- boolean offer(E element)

  adds the given element to the tail of this queue and returns true, provided the queue is not full. If the queue is full, the first method throws an IllegalStateException, whereas the second method returns false.

(Continues)

---

java.util.Queue<E> 5 *(Continued)*

- E remove()
- E poll()

removes and returns the element at the head of this queue, provided the queue is not empty. If the queue is empty, the first method throws a NoSuchElementException, whereas the second method returns null.

- E element()
- E peek()

returns the element at the head of this queue without removing it, provided the queue is not empty. If the queue is empty, the first method throws a NoSuchElementException, whereas the second method returns null.

---

java.util.Deque<E> 6

- void addFirst(E element)
- void addLast(E element)
- boolean offerFirst(E element)
- boolean offerLast(E element)

adds the given element to the head or tail of this deque. If the deque is full, the first two methods throw an IllegalStateException, whereas the last two methods return false.

- E removeFirst()
- E removeLast()
- E pollFirst()
- E pollLast()

removes and returns the element at the head of this deque, provided the deque is not empty. If the deque is empty, the first two methods throw a NoSuchElementException, whereas the last two methods return null.

- E getFirst()
- E getLast()
- E peekFirst()
- E peekLast()

returns the element at the head of this deque without removing it, provided the deque is not empty. If the deque is empty, the first two methods throw a NoSuchElementException, whereas the last two methods return null.

---

java.util.ArrayDeque<E> 6

- ArrayDeque()
- ArrayDeque(int initialCapacity)

  constructs an unbounded deque with an initial capacity of 16 or the given initial capacity.

9.3.6 Priority Queues

A priority queue retrieves elements in sorted order after they were inserted in arbitrary order. That is, whenever you call the remove method, you get the smallest element currently in the priority queue. However, the priority queue does not sort all its elements. If you iterate over the elements, they are not necessarily sorted. The priority queue makes use of an elegant and efficient data structure called a *heap*. A heap is a self-organizing binary tree in which the add and remove operations cause the smallest element to gravitate to the root, without wasting time on sorting all elements.

Just like a TreeSet, a priority queue can either hold elements of a class that implements the Comparable interface or a Comparator object you supply in the constructor.

A typical use for a priority queue is job scheduling. Each job has a priority. Jobs are added in random order. Whenever a new job can be started, the highest priority job is removed from the queue. (Since it is traditional for priority 1 to be the "highest" priority, the remove operation yields the minimum element.)

Listing 9.5 shows a priority queue in action. Unlike iteration in a TreeSet, the iteration here does not visit the elements in sorted order. However, removal always yields the smallest remaining element.

Listing 9.5 priorityQueue/PriorityQueueTest.java

```
1  package priorityQueue;
2
3  import java.util.*;
4  import java.time.*;
5
6  /**
7   * This program demonstrates the use of a priority queue.
8   * @version 1.02 2015-06-20
9   * @author Cay Horstmann
10  */
```

```
11  public class PriorityQueueTest
12  {
13     public static void main(String[] args)
14     {
15        var pq = new PriorityQueue<LocalDate>();
16        pq.add(LocalDate.of(1906, 12, 9)); // G. Hopper
17        pq.add(LocalDate.of(1815, 12, 10)); // A. Lovelace
18        pq.add(LocalDate.of(1903, 12, 3)); // J. von Neumann
19        pq.add(LocalDate.of(1910, 6, 22)); // K. Zuse
20
21        System.out.println("Iterating over elements . . .");
22        for (LocalDate date : pq)
23           System.out.println(date);
24        System.out.println("Removing elements . . .");
25        while (!pq.isEmpty())
26           System.out.println(pq.remove());
27     }
28  }
```

---

java.util.PriorityQueue 5

- PriorityQueue()
- PriorityQueue(int initialCapacity)

  constructs a priority queue for storing Comparable objects.

- PriorityQueue(int initialCapacity, Comparator<? super E> c)

  constructs a priority queue and uses the specified comparator for sorting its elements.

9.4 Maps

A set is a collection that lets you quickly find an existing element. However, to look up an element, you need to have an exact copy of the element to find. That isn't a very common lookup—usually, you have some key information, and you want to look up the associated element. The *map* data structure serves that purpose. A map stores key/value pairs. You can find a value if you provide the key. For example, you may store a table of employee records, where the keys are the employee IDs and the values are Employee objects. In the following sections, you will learn how to work with maps.

9.4.1 Basic Map Operations

The Java library supplies two general-purpose implementations for maps: HashMap and TreeMap. Both classes implement the Map interface.

A hash map hashes the keys, and a tree map uses an ordering on the keys to organize them in a search tree. The hash or comparison function is applied *only to the keys*. The values associated with the keys are not hashed or compared.

Should you choose a hash map or a tree map? As with sets, hashing is usually a bit faster, and it is the preferred choice if you don't need to visit the keys in sorted order.

Here is how you set up a hash map for storing employees:

```
var staff = new HashMap<String, Employee>(); // HashMap implements Map
var harry = new Employee("Harry Hacker");
staff.put("987-98-9996", harry);
. . .
```

Whenever you add an object to a map, you must supply a key as well. In our case, the key is a string, and the corresponding value is an Employee object.

To retrieve an object, you must use (and, therefore, remember) the key.

```
var id = "987-98-9996";
Employee e = staff.get(id); // gets harry
```

If no information is stored in the map with the particular key specified, get returns null.

The null return value can be inconvenient. Sometimes, you have a good default that can be used for keys that are not present in the map. Then use the getOrDefault method.

```
Map<String, Integer> scores = . . .;
int score = scores.getOrDefault(id, 0); // gets 0 if the id is not present
```

Keys must be unique. You cannot store two values with the same key. If you call the put method twice with the same key, the second value replaces the first one. In fact, put returns the previous value associated with its key parameter.

The remove method removes an element with a given key from the map. The size method returns the number of entries in the map.

The easiest way of iterating over the keys and values of a map is the forEach method. Provide a lambda expression that receives a key and a value. That expression is invoked for each map entry in turn.

```
scores.forEach((k, v) ->
    System.out.println("key=" + k + ", value=" + v));
```

Listing 9.6 illustrates a map at work. We first add key/value pairs to a map. Then, we remove one key from the map, which removes its associated value as well. Next, we change the value that is associated with a key and call the get method to look up a value. Finally, we iterate through the entry set.

Listing 9.6 map/MapTest.java

```java
 1 package map;
 2
 3 import java.util.*;
 4
 5 /**
 6  * This program demonstrates the use of a map with key type String and value type Employee.
 7  * @version 1.12 2015-06-21
 8  * @author Cay Horstmann
 9  */
10 public class MapTest
11 {
12    public static void main(String[] args)
13    {
14       var staff = new HashMap<String, Employee>();
15       staff.put("144-25-5464", new Employee("Amy Lee"));
16       staff.put("567-24-2546", new Employee("Harry Hacker"));
17       staff.put("157-62-7935", new Employee("Gary Cooper"));
18       staff.put("456-62-5527", new Employee("Francesca Cruz"));
19
20       // print all entries
21
22       System.out.println(staff);
23
24       // remove an entry
25
26       staff.remove("567-24-2546");
27
28       // replace an entry
29
30       staff.put("456-62-5527", new Employee("Francesca Miller"));
31
32       // look up a value
33
34       System.out.println(staff.get("157-62-7935"));
35
36       // iterate through all entries
37
```

(Continues)

Listing 9.6 *(Continued)*

```
38        staff.forEach((k, v) ->
39            System.out.println("key=" + k + ", value=" + v));
40    }
41 }
```

---

java.util.Map<K, V> 1.2

- V get(Object key)

  gets the value associated with the key; returns the object associated with the key, or null if the key is not found in the map. Implementing classes may forbid null keys.

- default V getOrDefault(Object key, V defaultValue)

  gets the value associated with the key; returns the object associated with the key, or defaultValue if the key is not found in the map.

- V put(K key, V value)

  puts the association of a key and a value into the map. If the key is already present, the new object replaces the old one previously associated with the key. This method returns the old value of the key, or null if the key was not previously present. Implementing classes may forbid null keys or values.

- void putAll(Map<? extends K, ? extends V> entries)

  adds all entries from the specified map to this map.

- boolean containsKey(Object key)

  returns true if the key is present in the map.

- boolean containsValue(Object value)

  returns true if the value is present in the map.

- default void forEach(BiConsumer<? super K,? super V> action) 8

  applies the action to all key/value pairs of this map.

---

java.util.HashMap<K, V> 1.2

- HashMap()
- HashMap(int initialCapacity)
- HashMap(int initialCapacity, float loadFactor)

  constructs an empty hash map with the specified capacity and load factor (a number between 0.0 and 1.0 that determines at what percentage of fullness the hash table will be rehashed into a larger one). The default load factor is 0.75.

> **java.util.TreeMap<K,V>** 1.2
>
> - TreeMap()
>
>   constructs an empty tree map for keys that implement the Comparable interface.
> - TreeMap(Comparator<? super K> c)
>
>   constructs a tree map and uses the specified comparator for sorting its keys.
> - TreeMap(Map<? extends K, ? extends V> entries)
>
>   constructs a tree map and adds all entries from a map.
> - TreeMap(SortedMap<? extends K, ? extends V> entries)
>
>   constructs a tree map, adds all entries from a sorted map, and uses the same element comparator as the given sorted map.

> **java.util.SortedMap<K, V>** 1.2
>
> - Comparator<? super K> comparator()
>
>   returns the comparator used for sorting the keys, or null if the keys are compared with the compareTo method of the Comparable interface.
> - K firstKey()
> - K lastKey()
>
>   returns the smallest or largest key in the map.

9.4.2 Updating Map Entries

A tricky part of dealing with maps is updating an entry. Normally, you get the old value associated with a key, update it, and put back the updated value. But you have to worry about the special case of the first occurrence of a key. Consider using a map for counting how often a word occurs in a file. When we see a word, we'd like to increment a counter like this:

```
counts.put(word, counts.get(word) + 1);
```

That works, except in the case when word is encountered for the first time. Then get returns null, and a NullPointerException occurs.

A simple remedy is to use the getOrDefault method:

```
counts.put(word, counts.getOrDefault(word, 0) + 1);
```

Another approach is to first call the putIfAbsent method. It only puts a value if the key was previously absent (or mapped to null).

```
counts.putIfAbsent(word, 0);
counts.put(word, counts.get(word) + 1); // now we know that get will succeed
```

But you can do better than that. The `merge` method simplifies this common operation. The call

```
counts.merge(word, 1, Integer::sum);
```

associates `word` with 1 if the key wasn't previously present, and otherwise combines the previous value and 1, using the `Integer::sum` function.

The API notes describe other methods for updating map entries that are less commonly used.

---

java.util.Map<K, V> 1.2

- `default V merge(K key, V value, BiFunction<? super V,? super V,? extends V> remappingFunction)` 8

  If key is associated with a non-null value v, applies the function to v and value and either associates key with the result or, if the result is null, removes the key. Otherwise, associates key with value. Returns get(key).

- `default V compute(K key, BiFunction<? super K,? super V,? extends V> remappingFunction)` 8

  Applies the function to key and get(key). Either associates key with the result or, if the result is null, removes the key. Returns get(key).

- `default V computeIfPresent(K key, BiFunction<? super K,? super V,? extends V> remappingFunction)` 8

  If key is associated with a non-null value v, applies the function to key and v and either associates key with the result or, if the result is null, removes the key. Returns get(key).

- `default V computeIfAbsent(K key, Function<? super K,? extends V> mappingFunction)` 8

  Applies the function to key unless key is associated with a non-null value. Either associates key with the result or, if the result is null, removes the key. Returns get(key).

- `default void replaceAll(BiFunction<? super K,? super V,? extends V> function)` 8

  Calls the function on all entries. Associates keys with non-null results and removes keys with null results.

- `default V putIfAbsent(K key, V value)` 8

  If key is absent or associated with null, associates it with value and returns null. Otherwise returns the associated value.

9.4.3 Map Views

The collections framework does not consider a map itself as a collection. (Other frameworks for data structures consider a map as a collection of key/value pairs, or as a collection of values indexed by the keys.) However, you can obtain *views* of the map—objects that implement the Collection interface or one of its subinterfaces.

There are three views: the set of keys, the collection of values (which is not a set), and the set of key/value pairs. The keys and key/value pairs form a set because there can be only one copy of a key in a map. The methods

```
Set<K> keySet()
Collection<V> values()
Set<Map.Entry<K, V>> entrySet()
```

return these three views. (The elements of the entry set are objects of a class implementing the Map.Entry interface.)

Note that the keySet is *not* a HashSet or TreeSet, but an object of some other class that implements the Set interface. The Set interface extends the Collection interface. Therefore, you can use a keySet as you would use any collection.

For example, you can enumerate all keys of a map:

```
Set<String> keys = map.keySet();
for (String key : keys)
{
    do something with key
}
```

If you want to look at both keys and values, you can avoid value lookups by enumerating the *entries*. Use the following code skeleton:

```
for (Map.Entry<String, Employee> entry : staff.entrySet())
{
    String k = entry.getKey();
    Employee v = entry.getValue();
    do something with k, v
}
```

 TIP: You can avoid the cumbersome Map.Entry by using a var declaration.

```
for (var entry : map.entrySet())
{
    do something with entry.getKey(), entry.getValue()
}
```

Or simply use the forEach method:

```
map.forEach((k, v) -> {
   do something with k, v
});
```

If you invoke the `remove` method of the iterator on the key set view, you actually remove the key *and its associated value* from the map. However, you cannot *add* an element to the key set view. It makes no sense to add a key without also adding a value. If you try to invoke the `add` method, it throws an `UnsupportedOperationException`. The entry set view has the same restriction, even though it would make conceptual sense to add a new key/value pair.

---

java.util.Map<K, V> 1.2

- `Set<Map.Entry<K, V>> entrySet()`

  returns a set view of `Map.Entry` objects, the key/value pairs in the map. You can remove elements from this set and they are removed from the map, but you cannot add any elements.

- `Set<K> keySet()`

  returns a set view of all keys in the map. You can remove elements from this set and the keys and associated values are removed from the map, but you cannot add any elements.

- `Collection<V> values()`

  returns a collection view of all values in the map. You can remove elements from this set and the removed value and its key are removed from the map, but you cannot add any elements.

---

java.util.Map.Entry<K, V> 1.2

- `K getKey()`
- `V getValue()`

  returns the key or value of this entry.

- `V setValue(V newValue)`

  changes the value *in the associated map* to the new value and returns the old value.

9.4.4 Weak Hash Maps

The collection class library has several map classes for specialized needs that we briefly discuss in this and the following sections.

The WeakHashMap class was designed to solve an interesting problem. What happens with a value whose key is no longer used anywhere in your program? Suppose the last reference to a key has gone away. Then, there is no longer any way to refer to the value object. But, as no part of the program has the key any more, the key/value pair cannot be removed from the map. Why can't the garbage collector remove it? Isn't it the job of the garbage collector to remove unused objects?

Unfortunately, it isn't quite so simple. The garbage collector traces *live* objects. As long as the map object is live, *all* buckets in it are live and won't be reclaimed. Thus, your program should take care to remove unused values from long-lived maps. Or, you can use a WeakHashMap instead. This data structure cooperates with the garbage collector to remove key/value pairs when the only reference to the key is the one from the hash table entry.

Here are the inner workings of this mechanism. The WeakHashMap uses *weak references* to hold keys. A WeakReference object holds a reference to another object—in our case, a hash table key. Objects of this type are treated in a special way by the garbage collector. Normally, if the garbage collector finds that a particular object has no references to it, it simply reclaims the object. However, if the object is reachable *only* by a WeakReference, the garbage collector still reclaims the object, but places the weak reference that led to it into a queue. The operations of the WeakHashMap periodically check that queue for newly arrived weak references. The arrival of a weak reference in the queue signifies that the key was no longer used by anyone and has been collected. The WeakHashMap then removes the associated entry.

9.4.5 Linked Hash Sets and Maps

The LinkedHashSet and LinkedHashMap classes remember in which order you inserted items. That way, you can avoid the seemingly random order of items in a hash table. As entries are inserted into the table, they are joined in a doubly linked list (see Figure 9.11).

For example, consider the following map insertions from Listing 9.6:

```
var staff = new LinkedHashMap<String, Employee>();
staff.put("144-25-5464", new Employee("Amy Lee"));
staff.put("567-24-2546", new Employee("Harry Hacker"));
staff.put("157-62-7935", new Employee("Gary Cooper"));
staff.put("456-62-5527", new Employee("Francesca Cruz"));
```

Then, staff.keySet().iterator() enumerates the keys in this order:

```
144-25-5464
567-24-2546
```

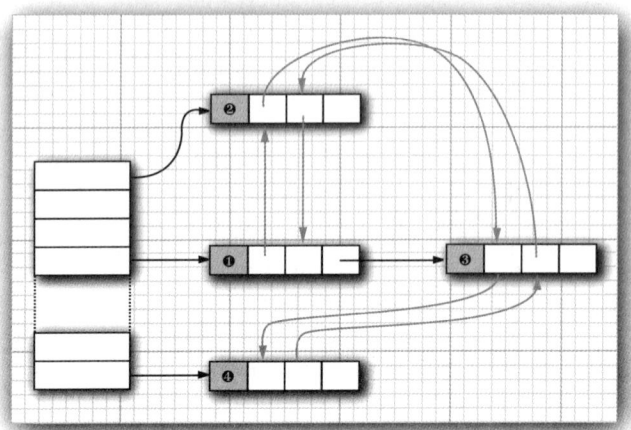

Figure 9.11 A linked hash table

```
157-62-7935
456-62-5527
```

and `staff.values().iterator()` enumerates the values in this order:

```
Amy Lee
Harry Hacker
Gary Cooper
Francesca Cruz
```

A linked hash map can alternatively use *access order*, not insertion order, to iterate through the map entries. Every time you call `get` or `put`, the affected entry is removed from its current position and placed at the *end* of the linked list of entries. (Only the position in the linked list of entries is affected, not the hash table bucket. An entry always stays in the bucket that corresponds to the hash code of the key.) To construct such a hash map, call

```
LinkedHashMap<K, V>(initialCapacity, loadFactor, true)
```

Access order is useful for implementing a "least recently used" discipline for a cache. For example, you may want to keep frequently accessed entries in memory and read less frequently accessed objects from a database. When you don't find an entry in the table, and the table is already pretty full, you can get an iterator into the table and remove the first few elements that it enumerates. Those entries were the least recently used ones.

You can even automate that process. Form a subclass of `LinkedHashMap` and override the method

```
protected boolean removeEldestEntry(Map.Entry<K, V> eldest)
```

Adding a new entry then causes the eldest entry to be removed whenever your method returns true. For example, the following cache is kept at a size of at most 100 elements:

```
var cache = new LinkedHashMap<K, V>(128, 0.75F, true)
   {
      protected boolean removeEldestEntry(Map.Entry<K, V> eldest)
      {
         return size() > 100;
      }
   };
```

Alternatively, you can consider the eldest entry to decide whether to remove it. For example, you may want to check a time stamp stored with the entry.

9.4.6 Enumeration Sets and Maps

The EnumSet is an efficient set implementation with elements that belong to an enumerated type. Since an enumerated type has a finite number of instances, the EnumSet is internally implemented simply as a sequence of bits. A bit is turned on if the corresponding value is present in the set.

The EnumSet class has no public constructors. Use a static factory method to construct the set:

```
enum Weekday { MONDAY, TUESDAY, WEDNESDAY, THURSDAY, FRIDAY, SATURDAY, SUNDAY };
EnumSet<Weekday> always = EnumSet.allOf(Weekday.class);
EnumSet<Weekday> never = EnumSet.noneOf(Weekday.class);
EnumSet<Weekday> workday = EnumSet.range(Weekday.MONDAY, Weekday.FRIDAY);
EnumSet<Weekday> mwf = EnumSet.of(Weekday.MONDAY, Weekday.WEDNESDAY, Weekday.FRIDAY);
```

You can use the usual methods of the Set interface to modify an EnumSet.

An EnumMap is a map with keys that belong to an enumerated type. It is simply and efficiently implemented as an array of values. You need to specify the key type in the constructor:

```
var personInCharge = new EnumMap<Weekday, Employee>(Weekday.class);
```

 NOTE: In the API documentation for EnumSet, you will see odd-looking type parameters of the form E extends Enum<E>. This simply means "E is an enumerated type." All enumerated types extend the generic Enum class. For example, Weekday extends Enum<Weekday>.

9.4.7 Identity Hash Maps

The IdentityHashMap has a quite specialized purpose. Here, the hash values for the keys should not be computed by the hashCode method but by the System.identityHashCode method. That's the method that Object.hashCode uses to compute a hash code from the object's memory address. Also, for comparison of objects, the IdentityHashMap uses ==, not equals.

In other words, different key objects are considered distinct even if they have equal contents. This class is useful for implementing object traversal algorithms, such as object serialization, in which you want to keep track of which objects have already been traversed.

---

java.util.WeakHashMap<K, V> 1.2

- WeakHashMap()
- WeakHashMap(int initialCapacity)
- WeakHashMap(int initialCapacity, float loadFactor)

  constructs an empty hash map with the specified capacity and load factor.

---

java.util.LinkedHashSet<E> 1.4

- LinkedHashSet()
- LinkedHashSet(int initialCapacity)
- LinkedHashSet(int initialCapacity, float loadFactor)

  constructs an empty linked hash set with the specified capacity and load factor.

---

java.util.LinkedHashMap<K, V> 1.4

- LinkedHashMap()
- LinkedHashMap(int initialCapacity)
- LinkedHashMap(int initialCapacity, float loadFactor)
- LinkedHashMap(int initialCapacity, float loadFactor, boolean accessOrder)

  constructs an empty linked hash map with the specified capacity, load factor, and ordering. The accessOrder parameter is true for access order, false for insertion order.

(Continues)

---

java.util.LinkedHashMap<K, V> 1.4 *(Continued)*

- protected boolean removeEldestEntry(Map.Entry<K, V> eldest)

  should be overridden to return `true` if you want the `eldest` entry to be removed. The `eldest` parameter is the entry whose removal is being contemplated. This method is called after an entry has been added to the map. The default implementation returns `false`—old elements are not removed by default. However, you can redefine this method to selectively return `true`—for example, if the eldest entry fits a certain condition or if the map exceeds a certain size.

---

java.util.EnumSet<E extends Enum<E>> 5

- static <E extends Enum<E>> EnumSet<E> allOf(Class<E> enumType)

  returns a mutable set that contains all values of the given enumerated type.

- static <E extends Enum<E>> EnumSet<E> noneOf(Class<E> enumType)

  returns a mutable set that is initially empty.

- static <E extends Enum<E>> EnumSet<E> range(E from, E to)

  returns a mutable set that contains all values between `from` and `to` (inclusive).

- static <E extends Enum<E>> EnumSet<E> of(E e)

  . . .

- static <E extends Enum<E>> EnumSet<E> of(E e1, E e2, E e3, E e4, E e5)
- static <E extends Enum<E>> EnumSet<E> of(E first, E... rest)

  returns a mutable set containing the given elements which must not be `null`.

---

java.util.EnumMap<K extends Enum<K>, V> 5

- EnumMap(Class<K> keyType)

  constructs an empty mutable map whose keys have the given type.

---

java.util.IdentityHashMap<K, V> 1.4

- IdentityHashMap()
- IdentityHashMap(int expectedMaxSize)

  constructs an empty identity hash map whose capacity is the smallest power of 2 exceeding 1.5 × expectedMaxSize. (The default for `expectedMaxSize` is 21.)

---

java.lang.System 1.0

- static int identityHashCode(Object obj) **1.1**

    returns the same hash code (derived from the object's memory address) that Object.hashCode computes, even if the class to which obj belongs has redefined the hashCode method.

9.5 Views and Wrappers

If you look at Figures 9.4 and 9.5, you might think it is overkill to have lots of interfaces and abstract classes to implement a modest number of concrete collection classes. However, these figures don't tell the whole story. By using *views*, you can obtain other objects that implement the Collection or Map interfaces. You saw one example of this with the keySet method of the map classes. At first glance, it appears as if the method creates a new set, fills it with all the keys of the map, and returns it. However, that is not the case. Instead, the keySet method returns an object of a class that implements the Set interface and whose methods manipulate the original map. Such a collection is called a *view*.

The technique of views has a number of useful applications in the collections framework. We will discuss these applications in the following sections.

9.5.1 Small Collections

Java 9 introduces static methods yielding a set or list with given elements, and a map with given key/value pairs.

For example,

```
List<String> names = List.of("Peter", "Paul", "Mary");
Set<Integer> numbers = Set.of(2, 3, 5);
```

yield a list and a set with three elements. For a map, you specify the keys and values, like this:

```
Map<String, Integer> scores = Map.of("Peter", 2, "Paul", 3, "Mary", 5);
```

The elements, keys, or values may not be null.

The List and Set interfaces have eleven of methods with zero to ten arguments, and an of method with a variable number of arguments. The specializations are provided for efficiency.

For the Map interface, it is not possible to provide a version with variable arguments since the argument types alternate between the key and value types. There is a static method ofEntries that accepts an arbitrary number of Map.Entry<K, V> objects, which you can create with the static entry method. For example,

```
import static java.util.Map.*;
. . .
Map<String, Integer> scores = ofEntries(
    entry("Peter", 2),
    entry("Paul", 3),
    entry("Mary", 5));
```

The of and ofEntries methods produce objects of classes that have an instance variable for each element, or that are backed by an array.

These collection objects are *unmodifiable*. Any attempt to change their contents results in an UnsupportedOperationException.

If you want a mutable collection, you can pass the unmodifiable collection to the constructor:

```
var names = new ArrayList<>(List.of("Peter", "Paul", "Mary"));
```

The method call

```
Collections.nCopies(n, anObject)
```

returns an immutable object that implements the List interface and gives the illusion of having n elements, each of which appears as anObject.

For example, the following call creates a List containing 100 strings, all set to "DEFAULT":

```
List<String> settings = Collections.nCopies(100, "DEFAULT");
```

There is very little storage cost—the object is stored only once.

NOTE: The of methods were introduced in Java 9. Previously, there was a static Arrays.asList method that returns a list that is mutable but not resizable. That is, you can call set but not add or remove on the list. There are also legacy methods Collections.emptySet and Collections.singleton.

NOTE: The Collections class contains a number of utility methods with parameters or return values that are collections. Do not confuse it with the Collection interface.

TIP: Java doesn't have a `Pair` class, and some programmers use a `Map.Entry` as a poor man's pair. Before Java 9, this was painful—you had to construct a `new AbstractMap.SimpleImmutableEntry<>(first, second)`. Nowadays, you can call `Map.entry(first, second)`.

9.5.2 Subranges

You can form subrange views for a number of collections. For example, suppose you have a list `staff` and want to extract elements 10 to 19. Use the `subList` method to obtain a view into the subrange of the list:

```
List<Employee> group2 = staff.subList(10, 20);
```

The first index is inclusive, the second exclusive—just like the parameters for the `substring` operation of the `String` class.

You can apply any operations to the subrange, and they automatically reflect the entire list. For example, you can erase the entire subrange:

```
group2.clear(); // staff reduction
```

The elements get automatically cleared from the `staff` list, and `group2` becomes empty.

For sorted sets and maps, you use the sort order, not the element position, to form subranges. The `SortedSet` interface declares three methods:

```
SortedSet<E> subSet(E from, E to)
SortedSet<E> headSet(E to)
SortedSet<E> tailSet(E from)
```

These return the subsets of all elements that are larger than or equal to `from` and strictly smaller than `to`. For sorted maps, the similar methods

```
SortedMap<K, V> subMap(K from, K to)
SortedMap<K, V> headMap(K to)
SortedMap<K, V> tailMap(K from)
```

return views into the maps consisting of all entries in which the *keys* fall into the specified ranges.

The `NavigableSet` interface introduced in Java 6 gives more control over these subrange operations. You can specify whether the bounds are included:

```
NavigableSet<E> subSet(E from, boolean fromInclusive, E to, boolean toInclusive)
NavigableSet<E> headSet(E to, boolean toInclusive)
NavigableSet<E> tailSet(E from, boolean fromInclusive)
```

9.5.3 Unmodifiable Views

The Collections class has methods that produce *unmodifiable views* of collections. These views add a runtime check to an existing collection. If an attempt to modify the collection is detected, an exception is thrown and the collection remains untouched.

You obtain unmodifiable views by eight methods:

```
Collections.unmodifiableCollection
Collections.unmodifiableList
Collections.unmodifiableSet
Collections.unmodifiableSortedSet
Collections.unmodifiableNavigableSet
Collections.unmodifiableMap
Collections.unmodifiableSortedMap
Collections.unmodifiableNavigableMap
```

Each method is defined to work on an interface. For example, Collections .unmodifiableList works with an ArrayList, a LinkedList, or any other class that implements the List interface.

For example, suppose you want to let some part of your code look at, but not touch, the contents of a collection. Here is what you could do:

```
var staff = new LinkedList<String>();
. . .
lookAt(Collections.unmodifiableList(staff));
```

The Collections.unmodifiableList method returns an object of a class implementing the List interface. Its accessor methods retrieve values from the staff collection. Of course, the lookAt method can call all methods of the List interface, not just the accessors. But all mutator methods (such as add) have been redefined to throw an UnsupportedOperationException instead of forwarding the call to the underlying collection.

The unmodifiable view does not make the collection itself immutable. You can still modify the collection through its original reference (staff, in our case). And you can still call mutator methods on the elements of the collection.

The views wrap the *interface* and not the actual collection object, so you only have access to those methods that are defined in the interface. For example, the LinkedList class has convenience methods, addFirst and addLast, that are not part of the List interface. These methods are not accessible through the unmodifiable view.

CAUTION: The `unmodifiableCollection` method (as well as the `synchronizedCollection` and `checkedCollection` methods discussed later in this section) returns a collection whose `equals` method does *not* invoke the `equals` method of the underlying collection. Instead, it inherits the `equals` method of the `Object` class, which just tests whether the objects are identical. If you turn a set or list into just a collection, you can no longer test for equal contents. The view acts in this way because equality testing is not well defined at this level of the hierarchy. The views treat the `hashCode` method in the same way.

However, the `unmodifiableSet` and `unmodifiableList` methods use the `equals` and `hashCode` methods of the underlying collections.

9.5.4 Synchronized Views

If you access a collection from multiple threads, you need to ensure that the collection is not accidentally damaged. For example, it would be disastrous if one thread tried to add to a hash table while another thread was rehashing the elements.

Instead of implementing thread-safe collection classes, the library designers used the view mechanism to make regular collections thread safe. For example, the static `synchronizedMap` method in the `Collections` class can turn any map into a `Map` with synchronized access methods:

```
var map = Collections.synchronizedMap(new HashMap<String, Employee>());
```

You can now access the `map` object from multiple threads. The methods such as `get` and `put` are synchronized—each method call must be finished completely before another thread can call another method. We discuss the issue of synchronized access to data structures in greater detail in Chapter 12.

9.5.5 Checked Views

Checked views are intended as debugging support for a problem that can occur with generic types. As explained in Chapter 8, it is actually possible to smuggle elements of the wrong type into a generic collection. For example:

```
var strings = new ArrayList<String>();
ArrayList rawList = strings; // warning only, not an error,
                             // for compatibility with legacy code
rawList.add(new Date()); // now strings contains a Date object!
```

The erroneous `add` command is not detected at runtime. Instead, a class cast exception will happen later when another part of the code calls `get` and casts the result to a `String`.

A checked view can detect this problem. Define a safe list as follows:

```
List<String> safeStrings = Collections.checkedList(strings, String.class);
```

The view's add method checks that the inserted object belongs to the given class and immediately throws a ClassCastException if it does not. The advantage is that the error is reported at the correct location:

```
ArrayList rawList = safeStrings;
rawList.add(new Date()); // checked list throws a ClassCastException
```

 CAUTION: The checked views are limited by the runtime checks that the virtual machine can carry out. For example, if you have an ArrayList<Pair<String>>, you cannot protect it from inserting a Pair<Date> since the virtual machine has a single "raw" Pair class.

9.5.6 A Note on Optional Operations

A view usually has some restriction—it may be read-only, it may not be able to change the size, or it may support removal but not insertion (as is the case for the key view of a map). A restricted view throws an UnsupportedOperationException if you attempt an inappropriate operation.

In the API documentation for the collection and iterator interfaces, many methods are described as "optional operations." This seems to be in conflict with the notion of an interface. After all, isn't the purpose of an interface to lay out the methods that a class *must* implement? Indeed, this arrangement is unsatisfactory from a theoretical perspective. A better solution might have been to design separate interfaces for read-only views and views that can't change the size of a collection. However, that would have tripled the number of interfaces, which the designers of the library found unacceptable.

Should you extend the technique of "optional" methods to your own designs? We think not. Even though collections are used frequently, the coding style for implementing them is not typical for other problem domains. The designers of a collection class library have to resolve a particularly brutal set of conflicting requirements. Users want the library to be easy to learn, convenient to use, completely generic, idiot-proof, and at the same time as efficient as hand-coded algorithms. It is plainly impossible to achieve all these goals simultaneously, or even to come close. But in your own programming problems, you will rarely encounter such an extreme set of constraints. You should be able to find solutions that do not rely on the extreme measure of "optional" interface operations.

---

java.util.List 1.2

- static <E> List<E> of() 9
- static <E> List<E> of(E el) 9

  . . .

- static <E> List<E> of(E el, E e2, E e3, E e4, E e5, E e6, E e7, E e8, E e9, E e10) 9
- static <E> Set<E> of(E... elements) 9

  yields an immutable list of the given elements, which must not be null.

---

java.util.Set 1.2

- static <E> Set<E> of() 9
- static <E> Set<E> of(E el) 9

  . . .

- static <E> Set<E> of(E el, E e2, E e3, E e4, E e5, E e6, E e7, E e8, E e9, E e10) 9
- static <E> Set<E> of(E... elements) 9

  yields an immutable set of the given elements, which must not be null.

---

java.util.Map 1.2

- static <K, V> Map<K, V> of() 9
- static <K, V> Map<K, V> of(K k1, V v1) 9

  . . .

- static <K,V> Map<K,V> of(K k1, V v1, K k2, V v2, K k3, V v3, K k4, V v4, K k5, V v5, K k6, V v6, K k7, V v7, K k8, V v8, K k9, V v9, K k10, V v10) 9

  yields an immutable map of the given keys and values, which must not be null.

- static <K,V> Map.Entry<K,V> entry(K k, V v) 9

  yields an immutable map entry of the given key and value, which must not be null.

- static <K,V> Map<K,V> ofEntries(Map.Entry<? extends K,? extends V>... entries) 9

  yields an immutable map of the given entries.

---

java.util.Collections 1.2

- static <E> Collection unmodifiableCollection(Collection<E> c)
- static <E> List unmodifiableList(List<E> c)
- static <E> Set unmodifiableSet(Set<E> c)
- static <E> SortedSet unmodifiableSortedSet(SortedSet<E> c)
- static <E> SortedSet unmodifiableNavigableSet(NavigableSet<E> c) 8
- static <K, V> Map unmodifiableMap(Map<K, V> c)
- static <K, V> SortedMap unmodifiableSortedMap(SortedMap<K, V> c)
- static <K, V> SortedMap unmodifiableNavigableMap(NavigableMap<K, V> c) 8

  constructs a view of the collection; the view's mutator methods throw an UnsupportedOperationException.

- static <E> Collection<E> synchronizedCollection(Collection<E> c)
- static <E> List synchronizedList(List<E> c)
- static <E> Set synchronizedSet(Set<E> c)
- static <E> SortedSet synchronizedSortedSet(SortedSet<E> c)
- static <E> NavigableSet synchronizedNavigableSet(NavigableSet<E> c) 8
- static <K, V> Map<K, V> synchronizedMap(Map<K, V> c)
- static <K, V> SortedMap<K, V> synchronizedSortedMap(SortedMap<K, V> c)
- static <K, V> NavigableMap<K, V> synchronizedNavigableMap(NavigableMap<K, V> c) 8

  constructs a view of the collection; the view's methods are synchronized.

- static <E> Collection checkedCollection(Collection<E> c, Class<E> elementType)
- static <E> List checkedList(List<E> c, Class<E> elementType)
- static <E> Set checkedSet(Set<E> c, Class<E> elementType)
- static <E> SortedSet checkedSortedSet(SortedSet<E> c, Class<E> elementType)
- static <E> NavigableSet checkedNavigableSet(NavigableSet<E> c, Class<E> elementType) 8
- static <K, V> Map checkedMap(Map<K, V> c, Class<K> keyType, Class<V> valueType)
- static <K, V> SortedMap checkedSortedMap(SortedMap<K, V> c, Class<K> keyType, Class<V> valueType)
- static <K, V> NavigableMap checkedNavigableMap(NavigableMap<K, V> c, Class<K> keyType, Class<V> valueType) 8
- static <E> Queue<E> checkedQueue(Queue<E> queue, Class<E> elementType) 8

  constructs a view of the collection; the view's methods throw a ClassCastException if an element of the wrong type is inserted.

- static <E> List<E> nCopies(int n, E value)

  yields an unmodifiable list with n identical values.

(Continues)

---

java.util.Collections 1.2 *(Continued)*

- static <E> List<E> singletonList(E value)
- static <E> Set<E> singleton(E value)
- static <K, V> Map<K, V> singletonMap(K key, V value)

yields a singleton list, set, or map. As of Java 9, use one of the of methods instead.

- static <E> List<E> emptyList()
- static <T> Set<T> emptySet()
- static <E> SortedSet<E> emptySortedSet()
- static NavigableSet<E> emptyNavigableSet()
- static <K,V> Map<K,V> emptyMap()
- static <K,V> SortedMap<K,V> emptySortedMap()
- static <K,V> NavigableMap<K,V> emptyNavigableMap()
- static <T> Enumeration<T> emptyEnumeration()
- static <T> Iterator<T> emptyIterator()
- static <T> ListIterator<T> emptyListIterator()

yields an empty collection, map, or iterator.

---

java.util.Arrays 1.2

- static <E> List<E> asList(E... array)

returns a list view of the elements in an array that is modifiable but not resizable.

---

java.util.List<E> 1.2

- List<E> subList(int firstIncluded, int firstExcluded)

returns a list view of the elements within a range of positions.

---

java.util.SortedSet<E> 1.2

- SortedSet<E> subSet(E firstIncluded, E firstExcluded)
- SortedSet<E> headSet(E firstExcluded)
- SortedSet<E> tailSet(E firstIncluded)

returns a view of the elements within a range.

---

java.util.NavigableSet<E> 6

- NavigableSet<E> subSet(E from, boolean fromIncluded, E to, boolean toIncluded)
- NavigableSet<E> headSet(E to, boolean toIncluded)
- NavigableSet<E> tailSet(E from, boolean fromIncluded)

  returns a view of the elements within a range. The boolean flags determine whether the bounds are included in the view.

---

java.util.SortedMap<K, V> 1.2

- SortedMap<K, V> subMap(K firstIncluded, K firstExcluded)
- SortedMap<K, V> headMap(K firstExcluded)
- SortedMap<K, V> tailMap(K firstIncluded)

  returns a map view of the entries whose keys are within a range.

---

java.util.NavigableMap<K, V> 6

- NavigableMap<K, V> subMap(K from, boolean fromIncluded, K to, boolean toIncluded)
- NavigableMap<K, V> headMap(K from, boolean fromIncluded)
- NavigableMap<K, V> tailMap(K to, boolean toIncluded)

  returns a map view of the entries whose keys are within a range. The boolean flags determine whether the bounds are included in the view.

9.6 Algorithms

In addition to implementing collection classes, the Java collections framework also provides a number of useful algorithms. In the following sections, you will see how to use these algorithms and how to write your own algorithms that work well with the collections framework.

9.6.1 Why Generic Algorithms?

Generic collection interfaces have a great advantage—you only need to implement your algorithms once. For example, consider a simple algorithm to compute the maximum element in a collection. Traditionally, programmers would implement such an algorithm as a loop. Here is how you can find the largest element of an array.

```
if (a.length == 0) throw new NoSuchElementException();
T largest = a[0];
```

```
for (int i = 1; i < a.length; i++)
   if (largest.compareTo(a[i]) < 0)
      largest = a[i];
```

Of course, to find the maximum of an array list, you would write the code slightly differently.

```
if (v.size() == 0) throw new NoSuchElementException();
T largest = v.get(0);
for (int i = 1; i < v.size(); i++)
   if (largest.compareTo(v.get(i)) < 0)
      largest = v.get(i);
```

What about a linked list? You don't have efficient random access in a linked list, but you can use an iterator.

```
if (l.isEmpty()) throw new NoSuchElementException();
Iterator<T> iter = l.iterator();
T largest = iter.next();
while (iter.hasNext())
{
   T next = iter.next();
   if (largest.compareTo(next) < 0)
      largest = next;
}
```

These loops are tedious to write, and just a bit error-prone. Is there an off-by-one error? Do the loops work correctly for empty containers? For containers with only one element? You don't want to test and debug this code every time, but you also don't want to implement a whole slew of methods, such as these:

```
static <T extends Comparable> T max(T[] a)
static <T extends Comparable> T max(ArrayList<T> v)
static <T extends Comparable> T max(LinkedList<T> l)
```

That's where the collection interfaces come in. Think of the *minimal* collection interface that you need to efficiently carry out the algorithm. Random access with get and set comes higher in the food chain than simple iteration. As you have seen in the computation of the maximum element in a linked list, random access is not required for this task. Computing the maximum can be done simply by iterating through the elements. Therefore, you can implement the max method to take *any* object that implements the Collection interface.

```
public static <T extends Comparable> T max(Collection<T> c)
{
   if (c.isEmpty()) throw new NoSuchElementException();
   Iterator<T> iter = c.iterator();
   T largest = iter.next();
```

```
   while (iter.hasNext())
   {
      T next = iter.next();
      if (largest.compareTo(next) < 0)
         largest = next;
   }
   return largest;
}
```

Now you can compute the maximum of a linked list, an array list, or an array, with a single method.

That's a powerful concept. In fact, the standard C++ library has dozens of useful algorithms, each operating on a generic collection. The Java library is not quite so rich, but it does contain the basics: sorting, binary search, and some utility algorithms.

9.6.2 Sorting and Shuffling

Computer old-timers will sometimes reminisce about how they had to use punched cards and to actually program, by hand, algorithms for sorting. Nowadays, of course, sorting algorithms are part of the standard library for most programming languages, and the Java programming language is no exception.

The sort method in the Collections class sorts a collection that implements the List interface.

```
var staff = new LinkedList<String>();
fill collection
Collections.sort(staff);
```

This method assumes that the list elements implement the Comparable interface. If you want to sort the list in some other way, you can use the sort method of the List interface and pass a Comparator object. Here is how you can sort a list of employees by salary:

```
staff.sort(Comparator.comparingDouble(Employee::getSalary));
```

If you want to sort a list in *descending* order, use the static convenience method Comparator.reverseOrder(). It returns a comparator that returns b.compareTo(a). For example,

```
staff.sort(Comparator.reverseOrder())
```

sorts the elements in the list staff in reverse order, according to the ordering given by the compareTo method of the element type. Similarly,

```
staff.sort(Comparator.comparingDouble(Employee::getSalary).reversed())
```

sorts by descending salary.

You may wonder how the sort method sorts a list. Typically, when you look at a sorting algorithm in a book on algorithms, it is presented for arrays and uses random element access. However, random access in a linked list is inefficient. You can actually sort linked lists efficiently by using a form of merge sort. However, the implementation in the Java programming language does not do that. It simply dumps all elements into an array, sorts the array, and then copies the sorted sequence back into the list.

The sort algorithm used in the collections library is a bit slower than Quick-Sort, the traditional choice for a general-purpose sorting algorithm. However, it has one major advantage: It is *stable*, that is, it doesn't switch equal elements. Why do you care about the order of equal elements? Here is a common scenario. Suppose you have an employee list that you already sorted by name. Now you sort by salary. What happens to employees with equal salary? With a stable sort, the ordering by name is preserved. In other words, the outcome is a list that is sorted first by salary, then by name.

Collections need not implement all of their "optional" methods, so all methods that receive collection parameters must describe when it is safe to pass a collection to an algorithm. For example, you clearly cannot pass an unmodifiableList list to the sort algorithm. What kind of list *can* you pass? According to the documentation, the list must be modifiable but need not be resizable.

The terms are defined as follows:

- A list is *modifiable* if it supports the set method.
- A list is *resizable* if it supports the add and remove operations.

The Collections class has an algorithm shuffle that does the opposite of sorting—it randomly permutes the order of the elements in a list. For example:

```
ArrayList<Card> cards = . . .;
Collections.shuffle(cards);
```

If you supply a list that does not implement the RandomAccess interface, the shuffle method copies the elements into an array, shuffles the array, and copies the shuffled elements back into the list.

The program in Listing 9.7 fills an array list with 49 Integer objects containing the numbers 1 through 49. It then randomly shuffles the list and selects the first six values from the shuffled list. Finally, it sorts the selected values and prints them.

Listing 9.7 shuffle/ShuffleTest.java

```
1  package shuffle;
2
3  import java.util.*;
4
5  /**
6   * This program demonstrates the random shuffle and sort algorithms.
7   * @version 1.12 2018-04-10
8   * @author Cay Horstmann
9   */
10 public class ShuffleTest
11 {
12    public static void main(String[] args)
13    {
14       var numbers = new ArrayList<Integer>();
15       for (int i = 1; i <= 49; i++)
16          numbers.add(i);
17       Collections.shuffle(numbers);
18       List<Integer> winningCombination = numbers.subList(0, 6);
19       Collections.sort(winningCombination);
20       System.out.println(winningCombination);
21    }
22 }
```

---

java.util.Collections 1.2

- static <T extends Comparable<? super T>> void sort(List<T> elements)

  sorts the elements in the list, using a stable sort algorithm. The algorithm is guaranteed to run in $O(n \log n)$ time, where n is the length of the list.

- static void shuffle(List<?> elements)
- static void shuffle(List<?> elements, Random r)

  randomly shuffles the elements in the list. This algorithm runs in $O(n\ a(n))$ time, where n is the length of the list and $a(n)$ is the average time to access an element.

---

java.util.List<E> 1.2

- default void sort(Comparator<? super T> comparator) 8

  Sorts this list, using the given comparator.

---

java.util.Comparator<T> 1.2

- static <T extends Comparable<? super T>> Comparator<T> reverseOrder() 8

  Yields a comparator that reverses the ordering provided by the Comparable interface.

- default Comparator<T> reversed() 8

  Yields a comparator that reverses the ordering provided by this comparator.

9.6.3 Binary Search

To find an object in an array, you normally visit all elements until you find a match. However, if the array is sorted, you can look at the middle element and check whether it is larger than the element that you are trying to find. If so, keep looking in the first half of the array; otherwise, look in the second half. That cuts the problem in half, and you keep going in the same way. For example, if the array has 1024 elements, you will locate the match (or confirm that there is none) after 10 steps, whereas a linear search would have taken you an average of 512 steps if the element is present, and 1024 steps to confirm that it is not.

The binarySearch of the Collections class implements this algorithm. Note that the collection must already be sorted, or the algorithm will return the wrong answer. To find an element, supply the collection (which must implement the List interface—more on that in the note below) and the element to be located. If the collection is not sorted by the compareTo element of the Comparable interface, supply a comparator object as well.

```
i = Collections.binarySearch(c, element);
i = Collections.binarySearch(c, element, comparator);
```

A non-negative return value from the binarySearch method denotes the index of the matching object. That is, c.get(i) is equal to element under the comparison order. If the value is negative, then there is no matching element. However, you can use the return value to compute the location where you *should* insert element into the collection to keep it sorted. The insertion location is

```
insertionPoint = -i - 1;
```

It isn't simply -i because then the value of 0 would be ambiguous. In other words, the operation

```
if (i < 0)
   c.add(-i - 1, element);
```

adds the element in the correct place.

To be worthwhile, binary search requires random access. If you have to iterate one by one through half of a linked list to find the middle element, you have lost all advantage of the binary search. Therefore, the `binarySearch` algorithm reverts to a linear search if you give it a linked list.

---

java.util.Collections 1.2

- `static <T extends Comparable<? super T>> int binarySearch(List<T> elements, T key)`
- `static <T> int binarySearch(List<T> elements, T key, Comparator<? super T> c)`

  searches for a key in a sorted list, using a binary search if the element type implements the `RandomAccess` interface, and a linear search in all other cases. The methods are guaranteed to run in $O(a(n) \log n)$ time, where n is the length of the list and $a(n)$ is the average time to access an element. The methods return either the index of the key in the list, or a negative value i if the key is not present in the list. In that case, the key should be inserted at index -i - 1 for the list to stay sorted.

9.6.4 Simple Algorithms

The `Collections` class contains several simple but useful algorithms. Among them is the example from the beginning of this section—finding the maximum value of a collection. Others include copying elements from one list to another, filling a container with a constant value, and reversing a list.

Why supply such simple algorithms in the standard library? Surely most programmers could easily implement them with simple loops. We like the algorithms because they make life easier for the programmer *reading* the code. When you read a loop that was implemented by someone else, you have to decipher the original programmer's intentions. For example, look at this loop:

```
for (int i = 0; i < words.size(); i++)
    if (words.get(i).equals("C++")) words.set(i, "Java");
```

Now compare the loop with the call

```
Collections.replaceAll(words, "C++", "Java");
```

When you see the method call, you know right away what the code does.

The API notes at the end of this section describe the simple algorithms in the `Collections` class.

The default methods `Collection.removeIf` and `List.replaceAll` that are just a bit more complex. You provide a lambda expression to test or transform elements. For example, here we remove all short words and change the remaining ones to lowercase:

```
words.removeIf(w -> w.length() <= 3);
words.replaceAll(String::toLowerCase);
```

---

java.util.Collections **1.2**

- `static <T extends Comparable<? super T>> T min(Collection<T> elements)`
- `static <T extends Comparable<? super T>> T max(Collection<T> elements)`
- `static <T> min(Collection<T> elements, Comparator<? super T> c)`
- `static <T> max(Collection<T> elements, Comparator<? super T> c)`

returns the smallest or largest element in the collection. (The parameter bounds are simplified for clarity.)

- `static <T> void copy(List<? super T> to, List<T> from)`

copies all elements from a source list to the same positions in the target list. The target list must be at least as long as the source list.

- `static <T> void fill(List<? super T> l, T value)`

sets all positions of a list to the same value.

- `static <T> boolean addAll(Collection<? super T> c, T... values)` **5**

adds all values to the given collection and returns `true` if the collection changed as a result.

- `static <T> boolean replaceAll(List<T> l, T oldValue, T newValue)` **1.4**

replaces all elements equal to `oldValue` with `newValue`.

- `static int indexOfSubList(List<?> l, List<?> s)` **1.4**
- `static int lastIndexOfSubList(List<?> l, List<?> s)` **1.4**

returns the index of the first or last sublist of `l` equaling `s`, or -1 if no sublist of `l` equals `s`. For example, if `l` is [s, t, a, r] and `s` is [t, a, r], then both methods return the index 1.

- `static void swap(List<?> l, int i, int j)` **1.4**

swaps the elements at the given offsets.

- `static void reverse(List<?> l)`

reverses the order of the elements in a list. For example, reversing the list [t, a, r] yields the list [r, a, t]. This method runs in $O(n)$ time, where n is the length of the list.

(Continues)

---

java.util.Collections 1.2 *(Continued)*

- static void rotate(List<?> l, int d) **1.4**

  rotates the elements in the list, moving the entry with index i to position (i + d) % l.size(). For example, rotating the list [t, a, r] by 2 yields the list [a, r, t]. This method runs in $O(n)$ time, where n is the length of the list.

- static int frequency(Collection<?> c, Object o) **5**

  returns the count of elements in c that equal the object o.

- boolean disjoint(Collection<?> c1, Collection<?> c2) **5**

  returns true if the collections have no elements in common.

---

java.util.Collection<T> 1.2

- default boolean removeIf(Predicate<? super E> filter) **8**

  removes all matching elements.

---

java.util.List<E> 1.2

- default void replaceAll(UnaryOperator<E> op) **8**

  applies the operation to all elements of this list.

9.6.5 Bulk Operations

There are several operations that copy or remove elements "in bulk." The call

```
coll1.removeAll(coll2);
```

removes all elements from coll1 that are present in coll2. Conversely,

```
coll1.retainAll(coll2);
```

removes all elements from coll1 that are *not* present in coll2. Here is a typical application.

Suppose you want to find the *intersection* of two sets—the elements that two sets have in common. First, make a new set to hold the result.

```
var result = new HashSet<String>(firstSet);
```

Here, we use the fact that every collection has a constructor whose parameter is another collection that holds the initialization values.

Now, use the `retainAll` method:

```
result.retainAll(secondSet);
```

It retains all elements that occur in both sets. You have formed the intersection without programming a loop.

You can carry this idea further and apply a bulk operation to a *view*. For example, suppose you have a map that maps employee IDs to employee objects and you have a set of the IDs of all employees that are to be terminated.

```
Map<String, Employee> staffMap = . . .;
Set<String> terminatedIDs = . . .;
```

Simply form the key set and remove all IDs of terminated employees.

```
staffMap.keySet().removeAll(terminatedIDs);
```

Since the key set is a view into the map, the keys and associated employee names are automatically removed from the map.

By using a subrange view, you can restrict bulk operations to sublists and subsets. For example, suppose you want to add the first ten elements of a list to another container. Form a sublist to pick out the first ten:

```
relocated.addAll(staff.subList(0, 10));
```

The subrange can also be a target of a mutating operation.

```
staff.subList(0, 10).clear();
```

9.6.6 Converting between Collections and Arrays

Large portions of the Java platform API were designed before the collections framework was created. As a result, you will occasionally need to translate between traditional arrays and the more modern collections.

If you have an array that you need to turn into a collection, the `List.of` wrapper serves this purpose. For example:

```
String[] values = . . .;
var staff = new HashSet<>(List.of(values));
```

Obtaining an array from a collection is a bit trickier. Of course, you can use the `toArray` method:

```
Object[] values = staff.toArray();
```

But the result is an array of *objects*. Even if you know that your collection contained objects of a specific type, you cannot use a cast:

```
String[] values = (String[]) staff.toArray(); // ERROR
```

The array returned by the toArray method was created as an Object[] array, and you cannot change its type. Instead, use a variant of the toArray method and give it an array of length 0 of the type that you'd like. The returned array is then created *as the same array type*:

```
String[] values = staff.toArray(new String[0]);
```

If you like, you can construct the array to have the correct size:

```
staff.toArray(new String[staff.size()]);
```

In this case, no new array is created.

 NOTE: You may wonder why you can't simply pass a Class object (such as String.class) to the toArray method. However, this method does "double duty"—both to fill an existing array (provided it is long enough) and to create a new array.

9.6.7 Writing Your Own Algorithms

If you write your own algorithm (or, in fact, any method that has a collection as a parameter), you should work with *interfaces*, not concrete implementations, whenever possible. For example, suppose you want to process items. Of course, you can implement a method like this:

```
public void processItems(ArrayList<Item> items)
{
   for (Item item : items)
      do something with item
}
```

However, you now constrained the caller of your method—the caller must supply the items in an ArrayList. If the items happen to be in another collection, they first need to be repackaged. It is much better to accept a more general collection.

You should ask yourself this: What is the most general collection interface that can do the job? Do you care about the order? Then you should accept a List. But if the order doesn't matter, you can accept collections of any kind:

```
public void processItems(Collection<Item> items)
{
    for (Item item : items)
        do something with item
}
```

Now, anyone can call this method with an ArrayList or a LinkedList, or even with an array wrapped with the List.of wrapper.

 TIP: In this case, you can do even better by accepting an Iterable<Item>. The Iterable interface has a single abstract method iterator which the enhanced for loop uses behind the scenes. The Collection interface extends Iterable.

Conversely, if your method returns multiple elements, you don't want to constrain yourself against future improvements. For example, consider

```
public ArrayList<Item> lookupItems(. . .)
{
    var result = new ArrayList<Item>();
    . . .
    return result;
}
```

This method promises to return an ArrayList, even though the caller almost certainly doesn't care what kind of lists it is. If instead you return a List, you can at any time add a branch that returns an empty or singleton list by calling List.of.

 NOTE: If it is such a good idea to use collection interfaces as parameter and return type, why doesn't the Java library follow this rule consistently? For example, the JComboBox class has two constructors:

```
JComboBox(Object[] items)
JComboBox(Vector<?> items)
```

The reason is simply timing. The Swing library was created before the collections library.

9.7 Legacy Collections

A number of "legacy" container classes have been present since the first release of Java, before there was a collections framework.

They have been integrated into the collections framework—see Figure 9.12. We briefly introduce them in the following sections.

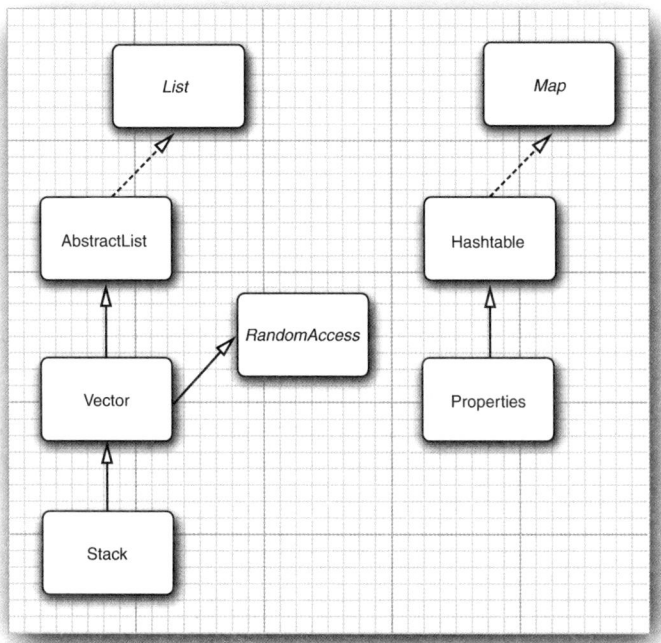

Figure 9.12 Legacy classes in the collections framework

9.7.1 The Hashtable Class

The classic Hashtable class serves the same purpose as the HashMap class and has essentially the same interface. Just like methods of the Vector class, the Hashtable methods are synchronized. If you do not require compatibility with legacy code, you should use a HashMap instead. If you need concurrent access, use a ConcurrentHashMap—see Chapter 12.

9.7.2 Enumerations

The legacy collections use the Enumeration interface for traversing sequences of elements. The Enumeration interface has two methods, hasMoreElements and nextElement. These are entirely analogous to the hasNext and next methods of the Iterator interface.

If you find this interface with legacy classes, you can use Collections.list to collect the elements in an ArrayList. For example, the LogManager class is only willing to reveal logger names as an Enumeration. Here is how you can get them all:

```
ArrayList<String> loggerNames = Collections.list(LogManager.getLoggerNames());
```

Alternatively, as of Java 9, you can turn an enumeration into an iterator:

```
LogManager.getLoggerNames().asIterator().forEachRemaining(n -> { . . . });
```

You will occasionally encounter a legacy method that expects an enumeration parameter. The static method Collections.enumeration yields an enumeration object that enumerates the elements in the collection. For example:

```
List<InputStream> streams = . . .;
var in = new SequenceInputStream(Collections.enumeration(streams));
    // the SequenceInputStream constructor expects an enumeration
```

 NOTE: In C++, it is quite common to use iterators as parameters. Fortunately, on the Java platform, very few programmers use this idiom. It is much smarter to pass around the collection than to pass an iterator. The collection object is more useful. The recipients can always obtain the iterator from the collection when they need to do so, plus they have all the collection methods at their disposal. However, you will find enumerations in some legacy code because they were the only available mechanism for generic collections until the collections framework appeared in Java 1.2.

---

java.util.Enumeration<E> 1.0

- boolean hasMoreElements()

  returns true if there are more elements yet to be inspected.

- E nextElement()

  returns the next element to be inspected. Do not call this method if hasMoreElements() returned false.

- default Iterator<E> asIterator() 9

  yields an iterator that iterates over the enumerated elements.

```
java.util.Collections  1.2
```

- static <T> Enumeration<T> enumeration(Collection<T> c)

  returns an enumeration that enumerates the elements of c.

- public static <T> ArrayList<T> list(Enumeration<T> e)

  returns an array list containing the elements enumerated by e.

9.7.3 Property Maps

A *property map* is a map structure of a special type. It has three particular characteristics:

- The keys and values are strings.
- The map can easily be saved to a file and loaded from a file.
- There is a secondary table for default values.

The Java platform class that implements a property map is called Properties. Property maps are useful in specifying configuration options for programs. For example:

```
var settings = new Properties();
settings.setProperty("width", "600.0");
settings.setProperty("filename", "/home/cay/books/cj11/code/v1ch11/raven.html");
```

Use the store method to save map list of properties to a file. Here, we just save the property map in the file program.properties. The second argument is a comment that is included in the file.

```
var out = new FileOutputStream("program.properties");
settings.store(out, "Program Properties");
```

The sample set gives the following output:

```
#Program Properties
#Sun Dec 31 12:54:19 PST 2017
top=227.0
left=1286.0
width=423.0
height=547.0
filename=/home/cay/books/cj11/code/v1ch11/raven.html
```

To load the properties from a file, use

```
var in = new FileInputStream("program.properties");
settings.load(in);
```

The System.getProperties method yields a Properties object to describe system information. For example, the home directory has the key "user.home". You can read it with the getProperties method that yields the key as a string:

```
String userDir = System.getProperty("user.home");
```

 CAUTION: For historical reasons, the Properties class implements Map<Object, Object>. Therefore, you can use the get and put methods of the Map interface. But the get method returns the type Object, and the put method allows you to insert any object. It is best to stick with the getProperty and setProperty methods that work with strings, not objects.

To get the Java version of the virtual machine, look up the "java.version" property. You get a string such as "9.0.1" (or "1.8.0" for Java 8.)

 TIP: As you can see, the version numbering changed in Java 9. This seemingly small change broke a good number of tools that had relied on the old format. If you parse the version string, be sure to read JEP 322 at http://openjdk.java.net/jeps/322 to see how version strings will be formatted in the future—or at least, until the numbering scheme changes again.

The Properties class has two mechanisms for providing defaults. First, whenever you look up the value of a string, you can specify a default that should be used automatically when the key is not present.

```
String filename = settings.getProperty("filename", "");
```

If there is a "filename" property in the property map, filename is set to that string. Otherwise, filename is set to the empty string.

If you find it too tedious to specify the default in every call to getProperty, you can pack all the defaults into a secondary property map and supply that map in the constructor of your primary property map.

```
var defaultSettings = new Properties();
defaultSettings.setProperty("width", "600");
defaultSettings.setProperty("height", "400");
defaultSettings.setProperty("filename", "");
. . .
var settings = new Properties(defaultSettings);
```

Yes, you can even specify defaults to defaults if you give another property map parameter to the defaultSettings constructor, but it is not something one would normally do.

The companion code has a sample program that shows how you can use properties for storing and loading program state. The program uses the ImageViewer program from Chapter 2 and remembers the frame position, size, and last loaded file. Run the program, load a file, and move and resize the window. Then close the program and reopen it to see that it remembers your file and your favorite window placement. You can also manually edit the file .corejava/ImageViewer.properties in your home directory.

 NOTE: Prior to Java 9, properties files used the 7-bit ASCII encoding. Nowadays, they use UTF-8.

Properties are simple tables without a hierarchical structure. It is common to introduce a fake hierarchy with key names such as window.main.color, window.main.title, and so on. But the Properties class has no methods that help organize such a hierarchy. If you store complex configuration information, you should use the Preferences class instead—see Chapter 10.

---

java.util.Properties 1.0

- Properties()

  creates an empty property map.

- Properties(Properties defaults)

  creates an empty property map with a set of defaults.

- String getProperty(String key)

  gets a property. Returns the string associated with the key, or the string associated with the key in the default table if it wasn't present in the table, or null if the key wasn't present in the default table either.

- String getProperty(String key, String defaultValue)

  gets a property with a default value if the key is not found. Returns the string associated with the key, or the default string if it wasn't present in the table.

- Object setProperty(String key, String value)

  sets a property. Returns the previously set value of the given key.

- void load(InputStream in) throws IOException

  loads a property map from an input stream.

- void store(OutputStream out, String header) 1.2

  saves a property map to an output stream. The header in the first line of the stored file.

`java.lang.System` 1.0

- `Properties getProperties()`

  retrieves all system properties. The application must have permission to retrieve all properties, or a security exception is thrown.

- `String getProperty(String key)`

  retrieves the system property with the given key name. The application must have permission to retrieve the property, or a security exception is thrown. The following properties can always be retrieved:

  ```
  java.version
  java.vendor
  java.vendor.url
  java.home
  java.class.path
  java.library.path
  java.class.version
  os.name
  os.version
  os.arch
  file.separator
  path.separator
  line.separator
  java.io.tempdir
  user.name
  user.home
  user.dir
  java.compiler
  java.specification.version
  java.specification.vendor
  java.specification.name
  java.vm.specification.version
  java.vm.specification.vendor
  java.vm.specification.name
  java.vm.version
  java.vm.vendor
  java.vm.name
  ```

9.7.4 Stacks

Since version 1.0, the standard library had a `Stack` class with the familiar `push` and `pop` methods. However, the `Stack` class extends the `Vector` class, which is not satisfactory from a theoretical perspective—you can apply such un-stack-like operations as `insert` and `remove` to insert and remove values anywhere, not just at the top of the stack.

---

java.util.Stack<E> 1.0

- E push(E item)

  pushes item onto the stack and returns item.

- E pop()

  pops and returns the top item of the stack. Don't call this method if the stack is empty.

- E peek()

  returns the top of the stack without popping it. Don't call this method if the stack is empty.

9.7.5 Bit Sets

The Java platform's BitSet class stores a sequence of bits. (It is not a *set* in the mathematical sense—bit *vector* or bit *array* would have been more appropriate terms.) Use a bit set if you need to store a sequence of bits (for example, flags) efficiently. A bit set packs the bits into bytes, so it is far more efficient to use a bit set than an ArrayList of Boolean objects.

The BitSet class gives you a convenient interface for reading, setting, and re-setting individual bits. Using this interface avoids the masking and other bit-fiddling operations that are necessary if you store bits in int or long variables.

For example, for a BitSet named bucketOfBits,

    bucketOfBits.get(i)

returns true if the ith bit is on, and false otherwise. Similarly,

    bucketOfBits.set(i)

turns the ith bit on. Finally,

    bucketOfBits.clear(i)

turns the ith bit off.

 C++ NOTE: The C++ bitset template has the same functionality as the Java platform BitSet.

---

java.util.BitSet 1.0

- BitSet(int initialCapacity)

  constructs a bit set.

- int length()

  returns the "logical length" of the bit set: 1 plus the index of the highest set bit.

- boolean get(int bit)

  gets a bit.

- void set(int bit)

  sets a bit.

- void clear(int bit)

  clears a bit.

- void and(BitSet set)

  logically ANDs this bit set with another.

- void or(BitSet set)

  logically ORs this bit set with another.

- void xor(BitSet set)

  logically XORs this bit set with another.

- void andNot(BitSet set)

  clears all bits in this bit set that are set in the other bit set.

As an example of using bit sets, we want to show you an implementation of the "sieve of Eratosthenes" algorithm for finding prime numbers. (A prime number is a number like 2, 3, or 5 that is divisible only by itself and 1, and the sieve of Eratosthenes was one of the first methods discovered to enumerate these fundamental building blocks.) This isn't a terribly good algorithm for finding the primes, but for some reason it has become a popular benchmark for compiler performance. (It isn't a good benchmark either, because it mainly tests bit operations.)

Oh well, we bow to tradition and present an implementation. This program counts all prime numbers between 2 and 2,000,000. (There are 148,933 primes in this interval, so you probably don't want to print them all out.)

Without going into too many details of this program, the idea is to march through a bit set with 2 million bits. First, we turn on all the bits. After that, we turn off the bits that are multiples of numbers known to be prime. The positions of the bits that remain after this process are themselves prime numbers. Listing 9.8 lists this program in the Java programming language, and Listing 9.9 is the C++ code.

 NOTE: Even though the sieve isn't a good benchmark, we couldn't resist timing the two implementations of the algorithm. Here are the timing results with a i7-8550U processor and 16 GB of RAM, running Ubuntu 17.10:

- C++ (g++ 7.2.0): 173 milliseconds
- Java (Java 9.0.1): 41 milliseconds

We have run this test for ten editions of *Core Java*, and in the last six editions, Java easily beat C++. In all fairness, if one cranks up the optimization level in the C++ compiler, it beats Java with a time of 34 milliseconds. Java could only match that if the program ran long enough to trigger the Hotspot just-in-time compiler.

Listing 9.8 sieve/Sieve.java

```java
1  package sieve;
2
3  import java.util.*;
4
5  /**
6   * This program runs the Sieve of Erathostenes benchmark. It computes all primes
7   * up to 2,000,000.
8   * @version 1.21 2004-08-03
9   * @author Cay Horstmann
10  */
11 public class Sieve
12 {
13    public static void main(String[] s)
14    {
15       int n = 2000000;
16       long start = System.currentTimeMillis();
17       var bitSet = new BitSet(n + 1);
18       int count = 0;
19       int i;
20       for (i = 2; i <= n; i++)
21          bitSet.set(i);
22       i = 2;
23       while (i * i <= n)
24       {
25          if (bitSet.get(i))
26          {
```

(Continues)

Listing 9.8 *(Continued)*

```
27            count++;
28            int k = 2 * i;
29            while (k <= n)
30            {
31               bitSet.clear(k);
32               k += i;
33            }
34         }
35         i++;
36      }
37      while (i <= n)
38      {
39         if (bitSet.get(i)) count++;
40         i++;
41      }
42      long end = System.currentTimeMillis();
43      System.out.println(count + " primes");
44      System.out.println((end - start) + " milliseconds");
45   }
46 }
```

Listing 9.9 sieve/sieve.cpp

```
1  /**
2     @version 1.21 2004-08-03
3     @author Cay Horstmann
4  */
5
6  #include <bitset>
7  #include <iostream>
8  #include <ctime>
9
10 using namespace std;
11
12 int main()
13 {
14    const int N = 2000000;
15    clock_t cstart = clock();
16
17    bitset<N + 1> b;
18    int count = 0;
19    int i;
20    for (i = 2; i <= N; i++)
21       b.set(i);
22    i = 2;
```

```
23      while (i * i <= N)
24      {
25         if (b.test(i))
26         {
27            count++;
28            int k = 2 * i;
29            while (k <= N)
30            {
31               b.reset(k);
32               k += i;
33            }
34         }
35         i++;
36      }
37      while (i <= N)
38      {
39         if (b.test(i))
40            count++;
41         i++;
42      }
43
44      clock_t cend = clock();
45      double millis = 1000.0 * (cend - cstart) / CLOCKS_PER_SEC;
46
47      cout << count << " primes\n" << millis << " milliseconds\n";
48
49      return 0;
50   }
```

This completes our tour through the Java collections framework. As you have seen, the Java library offers a wide variety of collection classes for your programming needs. In the next chapter, you will learn how to write graphical user interfaces.

CHAPTER **10**

Graphical User Interface Programming

In this chapter

Java was born at a time when most computer users interacted with graphical desktop applications. Nowadays, browser-based and mobile applications are far more common, but there are still times when it is useful to provide a desktop application. In this and the following chapter, we discuss the basics of user interface programming with the Swing toolkit. If, on the other hand, you intend to use Java for server-side programming only and are not interested in writing GUI programs, you can safely skip these two chapters.

10.1 A History of Java User Interface Toolkits

When Java 1.0 was introduced, it contained a class library, called the Abstract Window Toolkit (AWT), for basic GUI programming. The basic AWT library deals with user interface elements by delegating their creation and behavior to the native GUI toolkit on each target platform (Windows, Linux, Macintosh, and so on). For example, if you used the original AWT to put a text box on a Java window, an underlying "peer" text box actually handled the text input. The resulting program could then, in theory, run on any of these platforms, with the "look-and-feel" of the target platform.

The peer-based approach worked well for simple applications, but it soon became apparent that it was fiendishly difficult to write a high-quality portable graphics library depending on native user interface elements. User interface elements such as menus, scrollbars, and text fields can have subtle differences in behavior on different platforms. It was hard, therefore, to give users a consistent and predictable experience with this approach. Moreover, some graphical environments (such as X11/Motif) do not have as rich a collection of user interface components as does Windows or the Macintosh. This further limits a portable library based on a "lowest common denominator" approach. As a result, GUI applications built with the AWT simply did not look as nice as native Windows or Macintosh applications, nor did they have the kind of functionality that users of those platforms had come to expect. More depressingly, there were *different* bugs in the AWT user interface library on the different platforms. Developers complained that they had to test their applications on each platform—a practice derisively called "write once, debug everywhere."

In 1996, Netscape created a GUI library they called the IFC (Internet Foundation Classes) that used an entirely different approach. User interface elements, such as buttons, menus, and so on, were *painted* onto blank windows. The only functionality required from the underlying windowing system was a way to put up a window and to paint on it. Thus, Netscape's IFC widgets looked and behaved the same no matter which platform the program ran on. Sun Microsystems worked with Netscape to perfect this approach, creating a user interface library with the code name "Swing." Swing was available as an extension to Java 1.1 and became a part of the standard library in Java 1.2.

Swing is now the official name for the non-peer-based GUI toolkit.

 NOTE: Swing is not a complete replacement for the AWT—it is built on top of the AWT architecture. Swing simply gives you more capable user interface components. Whenever you write a Swing program, you use the foundations of the AWT—in particular, event handling. From now on, we say "Swing" when we mean the "painted" user interface classes, and we say "AWT" when we mean the underlying mechanisms of the windowing toolkit, such as event handling.

Swing has to work hard painting every pixel of the user interface. When Swing was first released, users complained that it was slow. (You can still get a feel for the problem if you run Swing applications on hardware such as a Raspberry Pi.) After a while, desktop computers got faster, and users complained that Swing was ugly—indeed, it had fallen behind the native widgets that had been spruced up with animations and fancy effects. More ominously, Adobe Flash was increasingly used to create user interfaces with even flashier effects that didn't use the native controls at all.

In 2007, Sun Microsystems introduced an entirely different user interface toolkit, called JavaFX, as a competitor to Flash. It ran on the Java VM but had its own programming language, called JavaFX Script. The language was optimized for programming animations and fancy effects. Programmers complained about the need to learn a new language, and they stayed away in droves. In 2011, Oracle released a new version, JavaFX 2.0, that had a Java API and no longer needed a separate programming language. Starting with Java 7 update 6, JavaFX has been bundled with the JDK and JRE. However, as this book is being written, Oracle has declared that JavaFX will no longer be bundled with Java, starting with version 11.

Since this is a book about the core Java language and APIs, we will focus on Swing for user interface programming.

 TIP: We provide you with a bonus chapter that introduces JavaFX. If you have a printed copy of this book, download a free PDF from `http://horstmann.com/corejava`. The ebook has the chapter at the end.

10.2 Displaying Frames

A top-level window (that is, a window that is not contained inside another window) is called a *frame* in Java. The AWT library has a class, called `Frame`, for this top level. The Swing version of this class is called `JFrame` and extends the `Frame` class. The `JFrame` is one of the few Swing components that is not painted on a canvas. Thus, the decorations (buttons, title bar, icons, and so on) are drawn by the user's windowing system, not by Swing.

 CAUTION: Most Swing component classes start with a "J": `JButton`, `JFrame`, and so on. There are classes such as `Button` and `Frame`, but they are AWT components. If you accidentally omit a "J", your program may still compile and run, but the mixture of Swing and AWT components can lead to visual and behavioral inconsistencies.

10.2.1 Creating a Frame

In this section, we will go over the most common methods for working with a Swing JFrame. Listing 10.1 lists a simple program that displays an empty frame on the screen, as illustrated in Figure 10.1.

Figure 10.1 The simplest visible frame

Listing 10.1 simpleframe/SimpleFrameTest.java

```
1  package simpleFrame;
2
3  import java.awt.*;
4  import javax.swing.*;
5
6  /**
7   * @version 1.34 2018-04-10
8   * @author Cay Horstmann
9   */
10 public class SimpleFrameTest
11 {
12    public static void main(String[] args)
13    {
14       EventQueue.invokeLater(() ->
15          {
16             var frame = new SimpleFrame();
17             frame.setDefaultCloseOperation(JFrame.EXIT_ON_CLOSE);
18             frame.setVisible(true);
19          });
20    }
21 }
22
```

```
23 class SimpleFrame extends JFrame
24 {
25    private static final int DEFAULT_WIDTH = 300;
26    private static final int DEFAULT_HEIGHT = 200;
27
28    public SimpleFrame()
29    {
30       setSize(DEFAULT_WIDTH, DEFAULT_HEIGHT);
31    }
32 }
```

Let's work through this program, line by line.

The Swing classes are placed in the javax.swing package. The package name javax indicates a Java extension package, not a core package. For historical reasons, Swing is considered an extension. However, it is present in every Java implementation since version 1.2.

By default, a frame has a rather useless size of 0 × 0 pixels. We define a subclass SimpleFrame whose constructor sets the size to 300 × 200 pixels. This is the only difference between a SimpleFrame and a JFrame.

In the main method of the SimpleFrameTest class, we construct a SimpleFrame object and make it visible.

There are two technical issues that we need to address in every Swing program.

First, all Swing components must be configured from the *event dispatch thread*, the thread of control that passes events such as mouse clicks and keystrokes to the user interface components. The following code fragment is used to execute statements in the event dispatch thread:

```
EventQueue.invokeLater(() ->
   {
      statements
   });
```

 NOTE: You will see many Swing programs that do not initialize the user interface in the event dispatch thread. It used to be perfectly acceptable to carry out the initialization in the main thread. Sadly, as Swing components got more complex, the developers of the JDK were no longer able to guarantee the safety of that approach. The probability of an error is extremely low, but you would not want to be one of the unlucky few who encounter an intermittent problem. It is better to do the right thing, even if the code looks rather mysterious.

Next, we define what should happen when the user closes the application's frame. For this particular program, we want the program to exit. To select this behavior, we use the statement

```
frame.setDefaultCloseOperation(JFrame.EXIT_ON_CLOSE);
```

In other programs with multiple frames, you would not want the program to exit just because the user closed one of the frames. By default, a frame is hidden when the user closes it, but the program does not terminate. (It might have been nice if the program terminated once the *last* frame becomes invisible, but that is not how Swing works.)

Simply constructing a frame does not automatically display it. Frames start their life invisible. That gives the programmer the chance to add components into the frame before showing it for the first time. To show the frame, the main method calls the setVisible method of the frame.

After scheduling the initialization statements, the main method exits. Note that exiting main does not terminate the program—just the main thread. The event dispatch thread keeps the program alive until it is terminated, either by closing the frame or by calling the System.exit method.

The running program is shown in Figure 10.1—it is a truly boring top-level window. As you can see in the figure, the title bar and the surrounding decorations, such as resize corners, are drawn by the operating system and not the Swing library. The Swing library draws everything inside the frame. In this program, it just fills the frame with a default background color.

10.2.2 Frame Properties

The JFrame class itself has only a few methods for changing how frames look. Of course, through the magic of inheritance, most of the methods for working with the size and position of a frame come from the various superclasses of JFrame. Here are some of the most important methods:

- The setLocation and setBounds methods for setting the position of the frame
- The setIconImage method, which tells the windowing system which icon to display in the title bar, task switcher window, and so on
- The setTitle method for changing the text in the title bar
- The setResizable method, which takes a boolean to determine if a frame will be resizeable by the user

Figure 10.2 illustrates the inheritance hierarchy for the JFrame class.

As the API notes indicate, the Component class (which is the ancestor of all GUI objects) and the Window class (which is the superclass of the Frame class) are

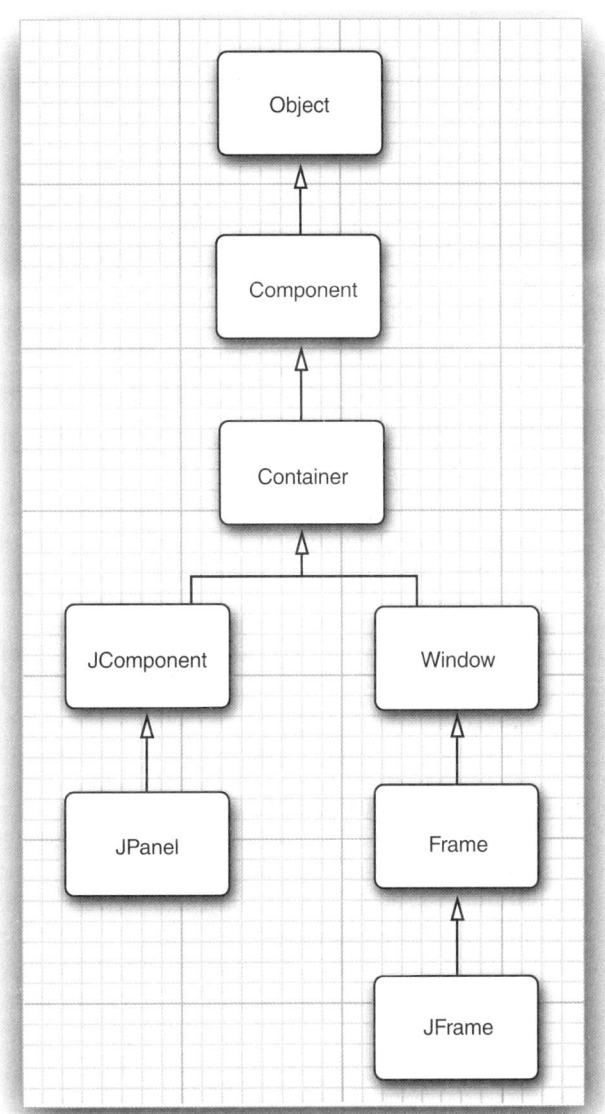

Figure 10.2 Inheritance hierarchy for the frame and component classes in AWT and Swing

where you need to look for the methods to resize and reshape frames. For example, the setLocation method in the Component class is one way to reposition a component. If you make the call

```
setLocation(x, y)
```

the top left corner is located x pixels across and y pixels down, where (0, 0) is the top left corner of the screen. Similarly, the setBounds method in Component lets you resize and relocate a component (in particular, a JFrame) in one step, as

```
setBounds(x, y, width, height)
```

Many methods of component classes come in getter/setter pairs, such as the following methods of the Frame class:

```
public String getTitle()
public void setTitle(String title)
```

Such a getter/setter pair is called a *property*. A property has a name and a type. The name is obtained by changing the first letter after the get or set to lowercase. For example, the Frame class has a property with name title and type String.

Conceptually, title is a property of the frame. When we set the property, we expect the title to change on the user's screen. When we get the property, we expect to get back the value that we have set.

There is one exception to the get/set convention: For properties of type boolean, the getter starts with is. For example, the following two methods define the resizable property:

```
public boolean isResizable()
public void setResizable(boolean resizable)
```

To determine an appropriate size for a frame, first find out the screen size. Call the static getDefaultToolkit method of the Toolkit class to get the Toolkit object. (The Toolkit class is a dumping ground for a variety of methods interfacing with the native windowing system.) Then call the getScreenSize method, which returns the screen size as a Dimension object. A Dimension object simultaneously stores a width and a height, in public (!) instance variables width and height. Then you can use a suitable percentage of the screen size to size the frame. Here is the code:

```
Toolkit kit = Toolkit.getDefaultToolkit();
Dimension screenSize = kit.getScreenSize();
int screenWidth = screenSize.width;
int screenHeight = screenSize.height;
setSize(screenWidth / 2, screenHeight / 2);
```

You can also supply frame icon:

```
Image img = new ImageIcon("icon.gif").getImage();
setIconImage(img);
```

---

java.awt.Component 1.0

- boolean isVisible()
- void setVisible(boolean b)

  gets or sets the visible property. Components are initially visible, with the exception of top-level components such as JFrame.

- void setSize(int width, int height) **1.1**

  resizes the component to the specified width and height.

- void setLocation(int x, int y) **1.1**

  moves the component to a new location. The x and y coordinates use the coordinates of the container if the component is not a top-level component, or the coordinates of the screen if the component is top level (for example, a JFrame).

- void setBounds(int x, int y, int width, int height) **1.1**

  moves and resizes this component.

- Dimension getSize() **1.1**
- void setSize(Dimension d) **1.1**

  gets or sets the size property of this component.

---

java.awt.Window 1.0

- void setLocationByPlatform(boolean b) **5**

  gets or sets the locationByPlatform property. When the property is set before this window is displayed, the platform picks a suitable location.

---

java.awt.Frame 1.0

- boolean isResizable()
- void setResizable(boolean b)

  gets or sets the resizable property. When the property is set, the user can resize the frame.

- String getTitle()
- void setTitle(String s)

  gets or sets the title property that determines the text in the title bar for the frame.

(Continues)

---

java.awt.Frame 1.0 *(Continued)*

- Image getIconImage()
- void setIconImage(Image image)

    gets or sets the iconImage property that determines the icon for the frame. The windowing system may display the icon as part of the frame decoration or in other locations.

---

java.awt.Toolkit 1.0

- static Toolkit getDefaultToolkit()

    returns the default toolkit.

- Dimension getScreenSize()

    gets the size of the user's screen.

---

javax.swing.ImageIcon 1.2

- ImageIcon(String filename)

    constructs an icon whose image is stored in a file.

- Image getImage()

    gets the image of this icon.

10.3 Displaying Information in a Component

In this section, we will show you how to display information inside a frame (Figure 10.3).

You could draw the message string directly onto a frame, but that is not considered good programming practice. In Java, frames are really designed to be containers for components, such as a menu bar and other user interface elements. You normally draw on another component which you add to the frame.

The structure of a JFrame is surprisingly complex. Look at Figure 10.4 which shows the makeup of a JFrame. As you can see, four panes are layered in a JFrame. The root pane, layered pane, and glass pane are of no interest to us; they are required to organize the menu bar and content pane and to implement the look-and-feel. The part that most concerns Swing programmers is

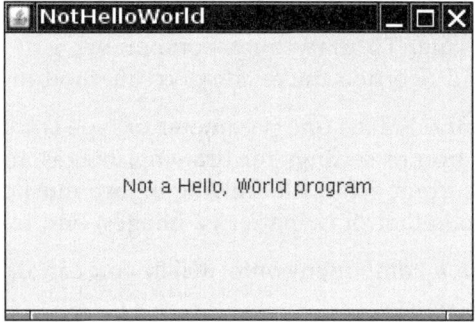

Figure 10.3 A frame that displays information

the *content pane*. Any components that you add to a frame are automatically placed into the content pane:

```
Component c = . . .;
frame.add(c); // added to the content pane
```

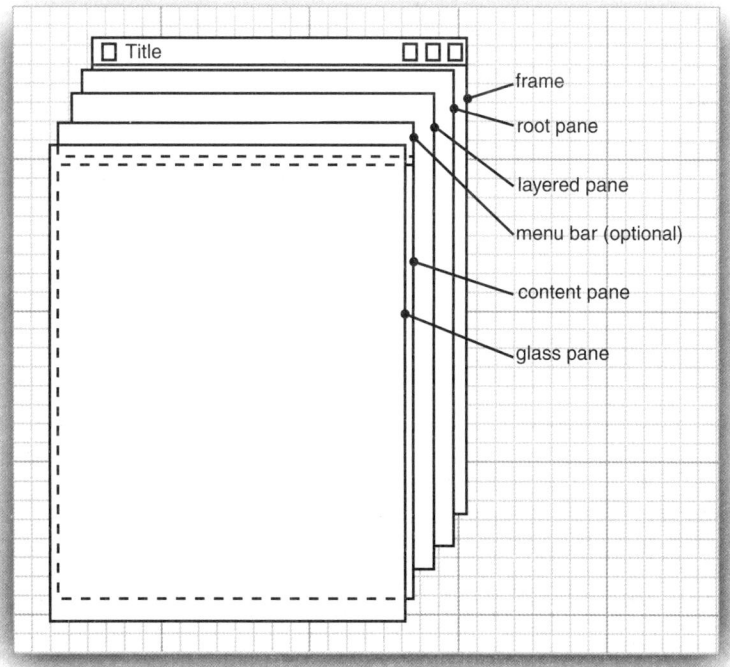

Figure 10.4 Internal structure of a JFrame

In our case, we want to add a single component to the frame onto which we will draw our message. To draw on a component, you define a class that extends JComponent and override the paintComponent method in that class.

The paintComponent method takes one parameter of type Graphics. A Graphics object remembers a collection of settings for drawing images and text, such as the font you set or the current color. All drawing in Java must go through a Graphics object. It has methods that draw patterns, images, and text.

Here's how to make a component onto which you can draw:

```
class MyComponent extends JComponent
{
   public void paintComponent(Graphics g)
   {
      code for drawing
   }
}
```

Each time a window needs to be redrawn, no matter what the reason, the event handler notifies the component. This causes the paintComponent methods of all components to be executed.

Never call the paintComponent method yourself. It is called automatically whenever a part of your application needs to be redrawn, and you should not interfere with this automatic process.

What sorts of actions trigger this automatic response? For example, painting occurs when the user increases the size of the window, or minimizes and then restores the window. If the user popped up another window that covered an existing window and then made the overlaid window disappear, the window that was covered is now corrupted and will need to be repainted. (The graphics system does not save the pixels underneath.) And, of course, when the window is displayed for the first time, it needs to process the code that specifies how and where it should draw the initial elements.

 TIP: If you need to force repainting of the screen, call the repaint method instead of paintComponent. The repaint method will cause paintComponent to be called for all components, with a properly configured Graphics object.

As you saw in the code fragment above, the paintComponent method takes a single parameter of type Graphics. Measurement on a Graphics object for screen display is done in pixels. The (0, 0) coordinate denotes the top left corner of the component on whose surface you are drawing.

The Graphics class has various drawing methods, and displaying text is considered a special kind of drawing. Our paintComponent method looks like this:

```
public class NotHelloWorldComponent extends JComponent
{
   public static final int MESSAGE_X = 75;
   public static final int MESSAGE_Y = 100;

   public void paintComponent(Graphics g)
   {
      g.drawString("Not a Hello, World program", MESSAGE_X, MESSAGE_Y);
   }
   . . .
}
```

Finally, a component should tell its users how big it would like to be. Override the getPreferredSize method and return an object of the Dimension class with the preferred width and height:

```
public class NotHelloWorldComponent extends JComponent
{
   private static final int DEFAULT_WIDTH = 300;
   private static final int DEFAULT_HEIGHT = 200;
   . . .
   public Dimension getPreferredSize()
   {
      return new Dimension(DEFAULT_WIDTH, DEFAULT_HEIGHT);
   }
}
```

When you fill a frame with one or more components, and you simply want to use their preferred size, call the pack method instead of the setSize method:

```
class NotHelloWorldFrame extends JFrame
{
   public NotHelloWorldFrame()
   {
      add(new NotHelloWorldComponent());
      pack();
   }
}
```

Listing 10.2 shows the complete code.

Listing 10.2 notHelloWorld/NotHelloWorld.java

```java
 1  package notHelloWorld;
 2
 3  import javax.swing.*;
 4  import java.awt.*;
 5
 6  /**
 7   * @version 1.34 2018-04-10
 8   * @author Cay Horstmann
 9   */
10  public class NotHelloWorld
11  {
12     public static void main(String[] args)
13     {
14        EventQueue.invokeLater(() ->
15           {
16              var frame = new NotHelloWorldFrame();
17              frame.setTitle("NotHelloWorld");
18              frame.setDefaultCloseOperation(JFrame.EXIT_ON_CLOSE);
19              frame.setVisible(true);
20           });
21     }
22  }
23
24  /**
25   * A frame that contains a message panel.
26   */
27  class NotHelloWorldFrame extends JFrame
28  {
29     public NotHelloWorldFrame()
30     {
31        add(new NotHelloWorldComponent());
32        pack();
33     }
34  }
35
36  /**
37   * A component that displays a message.
38   */
39  class NotHelloWorldComponent extends JComponent
40  {
41     public static final int MESSAGE_X = 75;
42     public static final int MESSAGE_Y = 100;
43
44     private static final int DEFAULT_WIDTH = 300;
45     private static final int DEFAULT_HEIGHT = 200;
46
```

```
47    public void paintComponent(Graphics g)
48    {
49       g.drawString("Not a Hello, World program", MESSAGE_X, MESSAGE_Y);
50    }
51
52    public Dimension getPreferredSize()
53    {
54       return new Dimension(DEFAULT_WIDTH, DEFAULT_HEIGHT);
55    }
56 }
```

---

javax.swing.JFrame 1.2

- Component add(Component c)

  adds and returns the given component to the content pane of this frame.

---

java.awt.Component 1.0

- void repaint()

  causes a repaint of the component "as soon as possible."

- Dimension getPreferredSize()

  is the method to override to return the preferred size of this component.

---

javax.swing.JComponent 1.2

- void paintComponent(Graphics g)

  is the method to override to describe how your component needs to be painted.

---

java.awt.Window 1.0

- void pack()

  resizes this window, taking into account the preferred sizes of its components.

10.3.1 Working with 2D Shapes

Starting with Java 1.0, the Graphics class has methods to draw lines, rectangles, ellipses, and so on. But those drawing operations are very limited. We will instead use the shape classes from the *Java 2D* library.

To use this library, you need to obtain an object of the Graphics2D class. This class is a subclass of the Graphics class. Ever since Java 1.2, methods such as paintComponent automatically receive an object of the Graphics2D class. Simply use a cast, as follows:

```
public void paintComponent(Graphics g)
{
   Graphics2D g2 = (Graphics2D) g;
   . . .
}
```

The Java 2D library organizes geometric shapes in an object-oriented fashion. In particular, there are classes to represent lines, rectangles, and ellipses:

```
Line2D
Rectangle2D
Ellipse2D
```

These classes all implement the Shape interface. The Java 2D library supports more complex shapes—arcs, quadratic and cubic curves, and general paths—that we do not discuss in this chapter.

To draw a shape, you first create an object of a class that implements the Shape interface and then call the draw method of the Graphics2D class. For example:

```
Rectangle2D rect = . . .;
g2.draw(rect);
```

The Java 2D library uses floating-point coordinates, not integers, for pixels. Internal calculations are carried out with single-precision float quantities. Single precision is sufficient—after all, the ultimate purpose of the geometric computations is to set pixels on the screen or printer. As long as any roundoff errors stay within one pixel, the visual outcome is not affected.

However, manipulating float values is sometimes inconvenient for the programmer because Java is adamant about requiring casts when converting double values into float values. For example, consider the following statement:

```
float f = 1.2; // ERROR--possible loss of precision
```

This statement does not compile because the constant 1.2 has type double, and the compiler is nervous about loss of precision. The remedy is to add an F suffix to the floating-point constant:

```
float f = 1.2F; // OK
```

Now consider this statement:

```
float f = r.getWidth(); // ERROR
```

This statement does not compile either, for the same reason. The getWidth method returns a double. This time, the remedy is to provide a cast:

```
float f = (float) r.getWidth(); // OK
```

These suffixes and casts are a bit of a pain, so the designers of the 2D library decided to supply *two versions* of each shape class: one with float coordinates for frugal programmers, and one with double coordinates for the lazy ones. (In this book, we fall into the second camp and use double coordinates whenever we can.)

The library designers chose a curious mechanism for packaging these choices. Consider the Rectangle2D class. This is an abstract class with two concrete subclasses, which are also static inner classes:

```
Rectangle2D.Float
Rectangle2D.Double
```

Figure 10.5 shows the inheritance diagram.

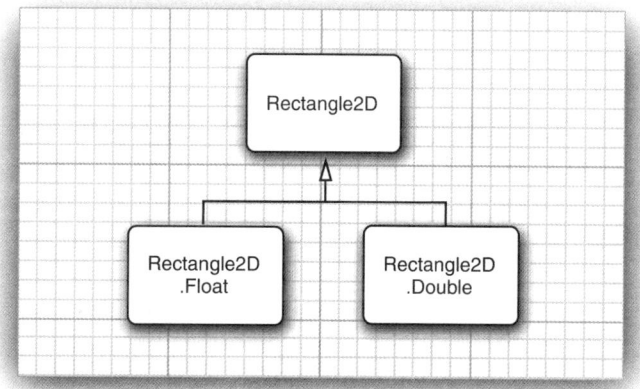

Figure 10.5 2D rectangle classes

It is best to ignore the fact that the two concrete classes are static inner classes—that is just a gimmick to avoid names such as FloatRectangle2D and DoubleRectangle2D.

When you construct a Rectangle2D.Float object, you supply the coordinates as float numbers. For a Rectangle2D.Double object, you supply them as double numbers.

```
var floatRect = new Rectangle2D.Float(10.0F, 25.0F, 22.5F, 20.0F);
var doubleRect = new Rectangle2D.Double(10.0, 25.0, 22.5, 20.0);
```

The construction parameters denote the top left corner, width, and height of the rectangle.

The Rectangle2D methods use double parameters and return values. For example, the getWidth method returns a double value, even if the width is stored as a float in a Rectangle2D.Float object.

 TIP: Simply use the Double shape classes to avoid dealing with float values altogether. However, if you are constructing thousands of shape objects, consider using the Float classes to conserve memory.

What we just discussed for the Rectangle2D classes holds for the other shape classes as well. Furthermore, there is a Point2D class with subclasses Point2D.Float and Point2D.Double. Here is how to make a point object:

```
var p = new Point2D.Double(10, 20);
```

The classes Rectangle2D and Ellipse2D both inherit from the common superclass RectangularShape. Admittedly, ellipses are not rectangular, but they have a *bounding rectangle* (see Figure 10.6).

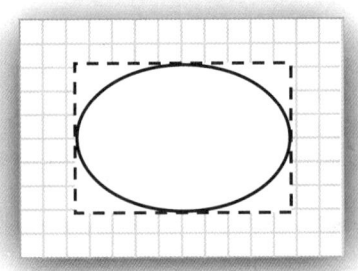

Figure 10.6 The bounding rectangle of an ellipse

The RectangularShape class defines over 20 methods that are common to these shapes, among them such useful methods as getWidth, getHeight, getCenterX, and getCenterY (but, sadly, at the time of this writing, not a getCenter method that would return the center as a Point2D object).

Finally, a couple of legacy classes from Java 1.0 have been fitted into the shape class hierarchy. The Rectangle and Point classes, which store a rectangle and a point with integer coordinates, extend the Rectangle2D and Point2D classes.

Figure 10.7 shows the relationships between the shape classes. However, the Double and Float subclasses are omitted. Legacy classes are marked with a gray fill.

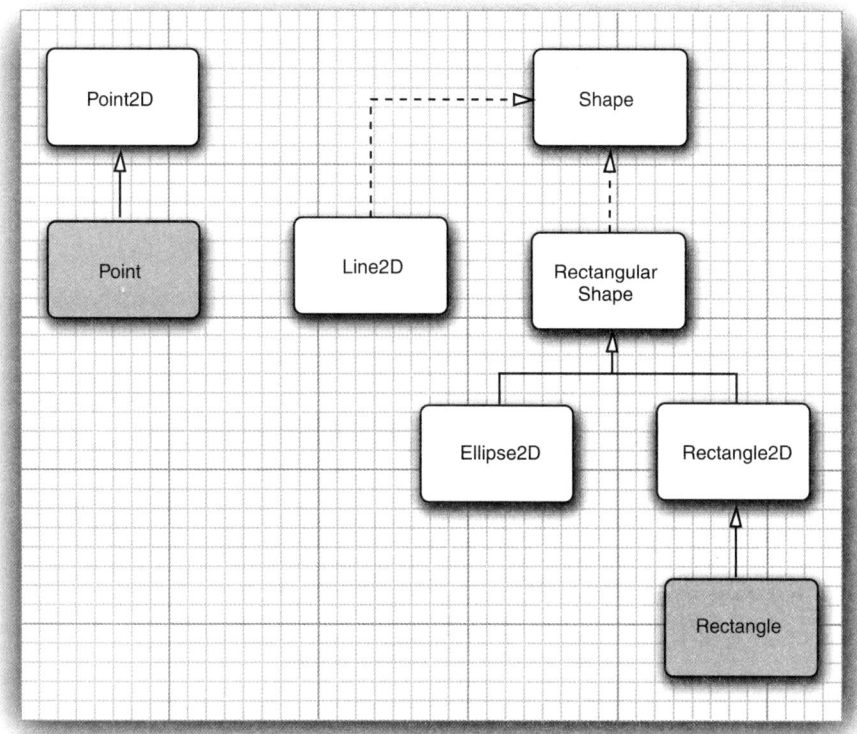

Figure 10.7 Relationships between the shape classes

Rectangle2D and Ellipse2D objects are simple to construct. You need to specify

- The x and y coordinates of the top left corner; and
- The width and height.

For ellipses, these refer to the bounding rectangle. For example,

```
var e = new Ellipse2D.Double(150, 200, 100, 50);
```

constructs an ellipse that is bounded by a rectangle with the top left corner at (150, 200), width of 100, and height of 50.

When constructing an ellipse, you usually know the center, width, and height, but not the corner points of the bounding rectangle (which don't even lie on the ellipse). The setFrameFromCenter method uses the center point, but it still requires one of the four corner points. Thus, you will usually end up constructing an ellipse as follows:

```
var ellipse
    = new Ellipse2D.Double(centerX - width / 2, centerY - height / 2, width, height);
```

To construct a line, you supply the start and end points, either as Point2D objects or as pairs of numbers:

```
var line = new Line2D.Double(start, end);
```

or

```
var line = new Line2D.Double(startX, startY, endX, endY);
```

The program in Listing 10.3 draws a rectangle, the ellipse that is enclosed in the rectangle, a diagonal of the rectangle, and a circle that has the same center as the rectangle. Figure 10.8 shows the result.

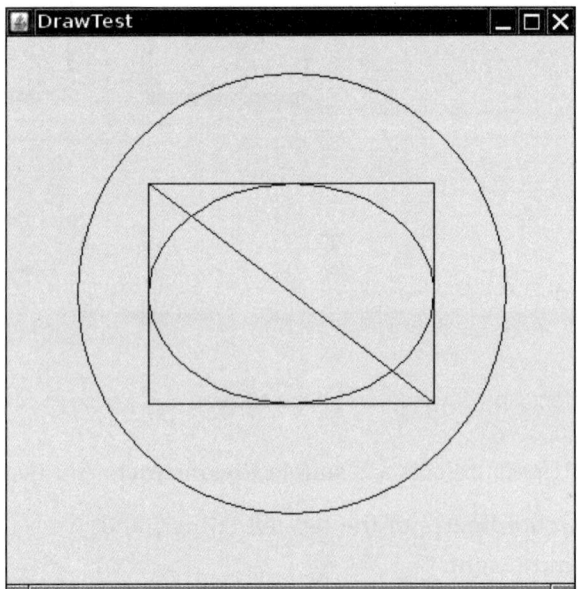

Figure 10.8 Drawing geometric shapes

Listing 10.3 draw/DrawTest.java

```
1 package draw;
2
3 import java.awt.*;
4 import java.awt.geom.*;
5 import javax.swing.*;
6
```

```
7  /**
8   * @version 1.34 2018-04-10
9   * @author Cay Horstmann
10  */
11 public class DrawTest
12 {
13    public static void main(String[] args)
14    {
15       EventQueue.invokeLater(() ->
16          {
17             var frame = new DrawFrame();
18             frame.setTitle("DrawTest");
19             frame.setDefaultCloseOperation(JFrame.EXIT_ON_CLOSE);
20             frame.setVisible(true);
21          });
22    }
23 }
24
25 /**
26  * A frame that contains a panel with drawings.
27  */
28 class DrawFrame extends JFrame
29 {
30    public DrawFrame()
31    {
32       add(new DrawComponent());
33       pack();
34    }
35 }
36
37 /**
38  * A component that displays rectangles and ellipses.
39  */
40 class DrawComponent extends JComponent
41 {
42    private static final int DEFAULT_WIDTH = 400;
43    private static final int DEFAULT_HEIGHT = 400;
44
45    public void paintComponent(Graphics g)
46    {
47       var g2 = (Graphics2D) g;
48
49       // draw a rectangle
50
51       double leftX = 100;
52       double topY = 100;
53       double width = 200;
54       double height = 150;
55
```

(Continues)

Listing 10.3 *(Continued)*

```
56        var rect = new Rectangle2D.Double(leftX, topY, width, height);
57        g2.draw(rect);
58
59        // draw the enclosed ellipse
60
61        var ellipse = new Ellipse2D.Double();
62        ellipse.setFrame(rect);
63        g2.draw(ellipse);
64
65        // draw a diagonal line
66
67        g2.draw(new Line2D.Double(leftX, topY, leftX + width, topY + height));
68
69        // draw a circle with the same center
70
71        double centerX = rect.getCenterX();
72        double centerY = rect.getCenterY();
73        double radius = 150;
74
75        var circle = new Ellipse2D.Double();
76        circle.setFrameFromCenter(centerX, centerY, centerX + radius, centerY + radius);
77        g2.draw(circle);
78    }
79
80    public Dimension getPreferredSize()
81    {
82        return new Dimension(DEFAULT_WIDTH, DEFAULT_HEIGHT);
83    }
84 }
```

---

java.awt.geom.RectangularShape 1.2

- double getCenterX()
- double getCenterY()
- double getMinX()
- double getMinY()
- double getMaxX()
- double getMaxY()

  returns the center, minimum, or maximum x or y value of the enclosing rectangle.

(Continues)

---

java.awt.geom.RectangularShape 1.2 *(Continued)*

- double getWidth()
- double getHeight()

  returns the width or height of the enclosing rectangle.

- double getX()
- double getY()

  returns the *x* or *y* coordinate of the top left corner of the enclosing rectangle.

---

java.awt.geom.Rectangle2D.Double 1.2

- Rectangle2D.Double(double x, double y, double w, double h)

  constructs a rectangle with the given top left corner, width, and height.

---

java.awt.geom.Ellipse2D.Double 1.2

- Ellipse2D.Double(double x, double y, double w, double h)

  constructs an ellipse whose bounding rectangle has the given top left corner, width, and height.

---

java.awt.geom.Point2D.Double 1.2

- Point2D.Double(double x, double y)

  constructs a point with the given coordinates.

---

java.awt.geom.Line2D.Double 1.2

- Line2D.Double(Point2D start, Point2D end)
- Line2D.Double(double startX, double startY, double endX, double endY)

  constructs a line with the given start and end points.

10.3.2 Using Color

The setPaint method of the Graphics2D class lets you select a color that is used for all subsequent drawing operations on the graphics context. For example:

```
g2.setPaint(Color.RED);
g2.drawString("Warning!", 100, 100);
```

You can fill the interiors of closed shapes (such as rectangles or ellipses) with a color. Simply call fill instead of draw:

```
Rectangle2D rect = . . .;
g2.setPaint(Color.RED);
g2.fill(rect); // fills rect with red
```

To draw in multiple colors, select a color, draw or fill, then select another color, and draw or fill again.

 NOTE: The fill method paints one fewer pixel to the right and the bottom. For example, if you draw a new Rectangle2D.Double(0, 0, 10, 20), then the drawing includes the pixels with $x = 10$ and $y = 20$. If you fill the same rectangle, those pixels are not painted.

Define colors with the Color class. The java.awt.Color class offers predefined constants for the following 13 standard colors:

```
BLACK, BLUE, CYAN, DARK_GRAY, GRAY, GREEN, LIGHT_GRAY,
MAGENTA, ORANGE, PINK, RED, WHITE, YELLOW
```

You can specify a custom color by creating a Color object by its red, green, and blue components, each a value between 0 and 255:

```
g2.setPaint(new Color(0, 128, 128)); // a dull blue-green
g2.drawString("Welcome!", 75, 125);
```

 NOTE: In addition to solid colors, you can call setPaint with instances of classes that implement the Paint interface. This enables drawing with gradients and textures.

To set the *background color,* use the setBackground method of the Component class, an ancestor of JComponent.

```
var component = new MyComponent();
component.setBackground(Color.PINK);
```

There is also a setForeground method. It specifies the default color that is used for drawing on the component.

---

java.awt.Color 1.0

- Color(int r, int g, int b)

  creates a color object with the given red, green, and blue components between 0 and 255.

---

java.awt.Graphics2D 1.2

- Paint getPaint()
- void setPaint(Paint p)

  gets or sets the paint property of this graphics context. The Color class implements the Paint interface. Therefore, you can use this method to set the paint attribute to a solid color.

- void fill(Shape s)

  fills the shape with the current paint.

---

java.awt.Component 1.0

- Color getForeground()
- Color getBackground()
- void setForeground(Color c)
- void setBackground(Color c)

  gets or sets the foreground or background color.

10.3.3 Using Fonts

The "Not a Hello World" program at the beginning of this chapter displayed a string in the default font. Sometimes, you will want to show your text in a different font. You can specify a font by its *font face name*. A font face name is composed of a *font family name*, such as "Helvetica", and an optional suffix such as "Bold". For example, the font faces "Helvetica" and "Helvetica Bold" are both considered to be part of the family named "Helvetica."

To find out which fonts are available on a particular computer, call the getAvailableFontFamilyNames method of the GraphicsEnvironment class. The method returns an array of strings containing the names of all available fonts. To obtain an instance of the GraphicsEnvironment class that describes the graphics environment of the user's system, use the static getLocalGraphicsEnvironment method. The following program prints the names of all fonts on your system:

```java
import java.awt.*;

public class ListFonts
{
   public static void main(String[] args)
   {
      String[] fontNames = GraphicsEnvironment
         .getLocalGraphicsEnvironment()
         .getAvailableFontFamilyNames();
      for (String fontName : fontNames)
         System.out.println(fontName);
   }
}
```

The AWT defines five *logical* font names:

```
SansSerif
Serif
Monospaced
Dialog
DialogInput
```

These names are always mapped to some fonts that actually exist on the client machine. For example, on a Windows system, SansSerif is mapped to Arial.

In addition, the Oracle JDK always includes three font families named "Lucida Sans," "Lucida Bright," and "Lucida Sans Typewriter."

To draw characters in a font, you must first create an object of the class Font. Specify the font face name, the font style, and the point size. Here is an example of how you construct a Font object:

```java
var sansbold14 = new Font("SansSerif", Font.BOLD, 14);
```

The third argument is the point size. Points are commonly used in typography to indicate the size of a font. There are 72 points per inch.

You can use a logical font name in place of the font face name in the Font constructor. Specify the style (plain, **bold**, *italic*, or ***bold italic***) by setting the second Font constructor argument to one of the following values:

```
Font.PLAIN
Font.BOLD
Font.ITALIC
Font.BOLD + Font.ITALIC
```

The font is plain with a font size of 1 point. Use the deriveFont method to get a font of the desired size:

```
Font f = f1.deriveFont(14.0F);
```

 CAUTION: There are two overloaded versions of the deriveFont method. One of them (with a float parameter) sets the font size, the other (with an int parameter) sets the font style. Thus, f1.deriveFont(14) sets the style and not the size! (The result is an italic font because it happens that the binary representation of 14 has the ITALIC bit but not the BOLD bit set.)

Here's the code that displays the string "Hello, World!" in the standard sans serif font on your system, using 14-point bold type:

```
var sansbold14 = new Font("SansSerif", Font.BOLD, 14);
g2.setFont(sansbold14);
var message = "Hello, World!";
g2.drawString(message, 75, 100);
```

Next, let's *center* the string in its component instead of drawing it at an arbitrary position. We need to know the width and height of the string in pixels. These dimensions depend on three factors:

- The font used (in our case, sans serif, bold, 14 point);
- The string (in our case, "Hello, World!"); and
- The device on which the font is drawn (in our case, the user's screen).

To obtain an object that represents the font characteristics of the screen device, call the getFontRenderContext method of the Graphics2D class. It returns an object of the FontRenderContext class. Simply pass that object to the getStringBounds method of the Font class:

```
FontRenderContext context = g2.getFontRenderContext();
Rectangle2D bounds = sansbold14.getStringBounds(message, context);
```

The getStringBounds method returns a rectangle that encloses the string.

To interpret the dimensions of that rectangle, you should know some basic typesetting terms (see Figure 10.9). The *baseline* is the imaginary line where, for example, the bottom of a character like 'e' rests. The *ascent* is the distance from the baseline to the top of an *ascender*, which is the upper part of a letter like 'b' or 'k', or an uppercase character. The *descent* is the distance from the baseline to a *descender*, which is the lower portion of a letter like 'p' or 'g'.

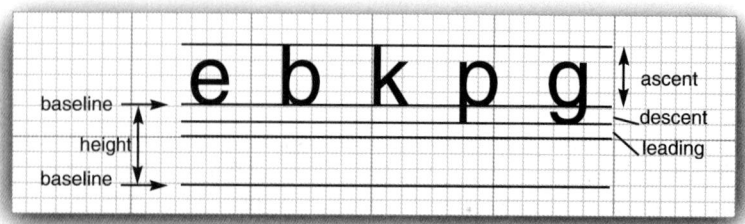

Figure 10.9 Typesetting terms illustrated

Leading is the space between the descent of one line and the ascent of the next line. (The term has its origin from the strips of lead that typesetters used to separate lines.) The *height* of a font is the distance between successive baselines, which is the same as descent + leading + ascent.

The width of the rectangle that the `getStringBounds` method returns is the horizontal extent of the string. The height of the rectangle is the sum of ascent, descent, and leading. The rectangle has its origin at the baseline of the string. The top *y* coordinate of the rectangle is negative. Thus, you can obtain string width, height, and ascent as follows:

```
double stringWidth = bounds.getWidth();
double stringHeight = bounds.getHeight();
double ascent = -bounds.getY();
```

If you need to know the descent or leading, use the `getLineMetrics` method of the `Font` class. That method returns an object of the `LineMetrics` class, which has methods to obtain the descent and leading:

```
LineMetrics metrics = f.getLineMetrics(message, context);
float descent = metrics.getDescent();
float leading = metrics.getLeading();
```

 NOTE: When you need to compute layout dimensions outside the `paintComponent` method, you can't obtain the font render context from the `Graphics2D` object. Instead, call the `getFontMetrics` method of the `JComponent` class and then call `getFontRenderContext`.

```
FontRenderContext context = getFontMetrics(f).getFontRenderContext();
```

To show that the positioning is accurate, the sample program in Listing 10.4 centers the string in the frame and draws the baseline and the bounding rectangle. Figure 10.10 shows the screen display.

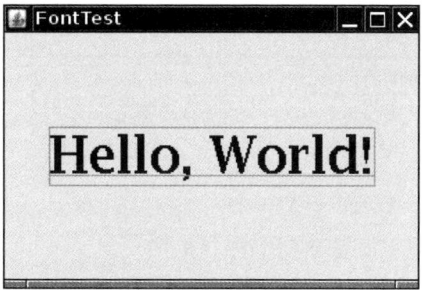

Figure 10.10 Drawing the baseline and string bounds

Listing 10.4 font/FontTest.java

```java
1  package font;
2
3  import java.awt.*;
4  import java.awt.font.*;
5  import java.awt.geom.*;
6  import javax.swing.*;
7
8  /**
9   * @version 1.35 2018-04-10
10  * @author Cay Horstmann
11  */
12 public class FontTest
13 {
14    public static void main(String[] args)
15    {
16       EventQueue.invokeLater(() ->
17          {
18             var frame = new FontFrame();
19             frame.setTitle("FontTest");
20             frame.setDefaultCloseOperation(JFrame.EXIT_ON_CLOSE);
21             frame.setVisible(true);
22          });
23    }
24 }
25
26 /**
27  * A frame with a text message component.
28  */
29 class FontFrame extends JFrame
30 {
31    public FontFrame()
32    {
```

(Continues)

Listing 10.4 *(Continued)*

```
33        add(new FontComponent());
34        pack();
35     }
36 }
37
38 /**
39  * A component that shows a centered message in a box.
40  */
41 class FontComponent extends JComponent
42 {
43    private static final int DEFAULT_WIDTH = 300;
44    private static final int DEFAULT_HEIGHT = 200;
45
46    public void paintComponent(Graphics g)
47    {
48       var g2 = (Graphics2D) g;
49
50       var message = "Hello, World!";
51
52       var f = new Font("Serif", Font.BOLD, 36);
53       g2.setFont(f);
54
55       // measure the size of the message
56
57       FontRenderContext context = g2.getFontRenderContext();
58       Rectangle2D bounds = f.getStringBounds(message, context);
59
60       // set (x,y) = top left corner of text
61
62       double x = (getWidth() - bounds.getWidth()) / 2;
63       double y = (getHeight() - bounds.getHeight()) / 2;
64
65       // add ascent to y to reach the baseline
66
67       double ascent = -bounds.getY();
68       double baseY = y + ascent;
69
70       // draw the message
71
72       g2.drawString(message, (int) x, (int) baseY);
73
74       g2.setPaint(Color.LIGHT_GRAY);
75
76       // draw the baseline
77
78       g2.draw(new Line2D.Double(x, baseY, x + bounds.getWidth(), baseY));
```

```
79
80      // draw the enclosing rectangle
81
82      var rect = new Rectangle2D.Double(x, y, bounds.getWidth(), bounds.getHeight());
83      g2.draw(rect);
84   }
85
86   public Dimension getPreferredSize()
87   {
88      return new Dimension(DEFAULT_WIDTH, DEFAULT_HEIGHT);
89   }
90 }
```

---

java.awt.Font 1.0

- `Font(String name, int style, int size)`

  creates a new font object. The font name is either a font face name (such as "Helvetica Bold") or a logical font name (such as "Serif", "SansSerif"). The style is one of Font.PLAIN, Font.BOLD, Font.ITALIC, or Font.BOLD + Font.ITALIC.

- `String getFontName()`

  gets the font face name (such as "Helvetica Bold").

- `String getFamily()`

  gets the font family name (such as "Helvetica").

- `String getName()`

  gets the logical name (such as "SansSerif") if the font was created with a logical font name; otherwise, gets the font face name.

- `Rectangle2D getStringBounds(String s, FontRenderContext context)` 1.2

  returns a rectangle that encloses the string. The origin of the rectangle falls on the baseline. The top y coordinate of the rectangle equals the negative of the ascent. The height of the rectangle equals the sum of ascent, descent, and leading. The width equals the string width.

- `LineMetrics getLineMetrics(String s, FontRenderContext context)` 1.2

  returns a line metrics object to determine the extent of the string.

- `Font deriveFont(int style)` 1.2
- `Font deriveFont(float size)` 1.2
- `Font deriveFont(int style, float size)` 1.2

  returns a new font that is equal to this font, except that it has the given size and style.

---

java.awt.font.LineMetrics 1.2

- float getAscent()

  gets the font ascent—the distance from the baseline to the tops of uppercase characters.

- float getDescent()

  gets the font descent—the distance from the baseline to the bottoms of descenders.

- float getLeading()

  gets the font leading—the space between the bottom of one line of text and the top of the next line.

- float getHeight()

  gets the total height of the font—the distance between the two baselines of text (descent + leading + ascent).

---

java.awt.Graphics2D 1.2

- FontRenderContext getFontRenderContext()

  gets a font render context that specifies font characteristics in this graphics context.

- void drawString(String str, float x, float y)

  draws a string in the current font and color.

---

javax.swing.JComponent 1.2

- FontMetrics getFontMetrics(Font f) **5**

  gets the font metrics for the given font. The FontMetrics class is a precursor to the LineMetrics class.

---

java.awt.FontMetrics 1.0

- FontRenderContext getFontRenderContext() **1.2**

  gets a font render context for the font.

10.3.4 Displaying Images

You can use the `ImageIcon` class to read an image from a file:

```
Image image = new ImageIcon(filename).getImage();
```

Now the variable `image` contains a reference to an object that encapsulates the image data. Display the image with the `drawImage` method of the `Graphics` class.

```
public void paintComponent(Graphics g)
{
   . . .
   g.drawImage(image, x, y, null);
}
```

We can take this a little bit further and tile the window with the graphics image. The result looks like the screen shown in Figure 10.11. We do the tiling in the `paintComponent` method. We first draw one copy of the image in the top left corner and then use the `copyArea` call to copy it into the entire window:

```
for (int i = 0; i * imageWidth <= getWidth(); i++)
   for (int j = 0; j * imageHeight <= getHeight(); j++)
      if (i + j > 0)
         g.copyArea(0, 0, imageWidth, imageHeight, i * imageWidth, j * imageHeight);
```

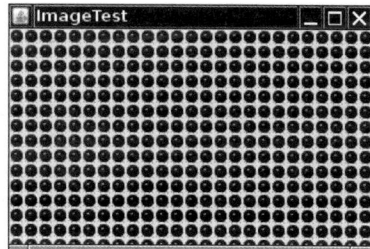

Figure 10.11 Window with tiled graphics image

`java.awt.Graphics` 1.0

- `boolean drawImage(Image img, int x, int y, ImageObserver observer)`
- `boolean drawImage(Image img, int x, int y, int width, int height, ImageObserver observer)`

  draws an unscaled or scaled image. Note: This call may return before the image is drawn. The `imageObserver` object is notified of the rendering progress. This was a useful feature in the distant past. Nowadays, just pass a `null` observer.

- `void copyArea(int x, int y, int width, int height, int dx, int dy)`

  copies an area of the screen. The `dx` and `dy` parameters are the distance from the source area to the target area.

10.4 Event Handling

Any operating environment that supports GUIs constantly monitors events such as keystrokes or mouse clicks. These events are then reported to the programs that are running. Each program then decides what, if anything, to do in response to these events.

10.4.1 Basic Event Handling Concepts

In the Java AWT, *event sources* (such as buttons or scrollbars) have methods that allow you to register *event listeners*—objects that carry out the desired response to the event.

When an event listener is notified about an event, information about the event is encapsulated in an *event object*. In Java, all event objects ultimately derive from the class java.util.EventObject. Of course, there are subclasses for each event type, such as ActionEvent and WindowEvent.

Different event sources can produce different kinds of events. For example, a button can send ActionEvent objects, whereas a window can send WindowEvent objects.

To sum up, here's an overview of how event handling in the AWT works:

* An event listener is an instance of a class that implements a *listener interface*.
* An event source is an object that can register listener objects and send them event objects.
* The event source sends out event objects to all registered listeners when that event occurs.
* The listener objects then uses the information in the event object to determine their reaction to the event.

Figure 10.12 shows the relationship between the event handling classes and interfaces.

Here is an example for specifying a listener:

```
ActionListener listener = . . .;
var button = new JButton("OK");
button.addActionListener(listener);
```

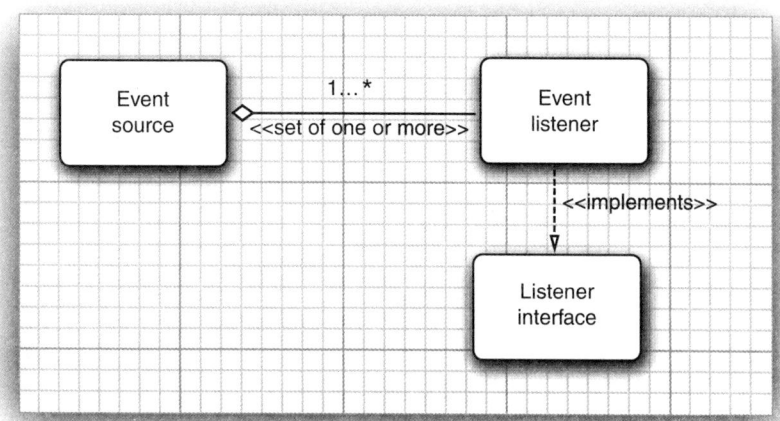

Figure 10.12 Relationship between event sources and listeners

Now the listener object is notified whenever an "action event" occurs in the button. For buttons, as you might expect, an action event is a button click.

To implement the ActionListener interface, the listener class must have a method called actionPerformed that receives an ActionEvent object as a parameter.

```
class MyListener implements ActionListener
{
   . . .
   public void actionPerformed(ActionEvent event)
   {
      // reaction to button click goes here
      . . .
   }
}
```

Whenever the user clicks the button, the JButton object creates an ActionEvent object and calls listener.actionPerformed(event), passing that event object. An event source such as a button can have multiple listeners. In that case, the button calls the actionPerformed methods of all listeners whenever the user clicks the button.

Figure 10.13 shows the interaction between the event source, event listener, and event object.

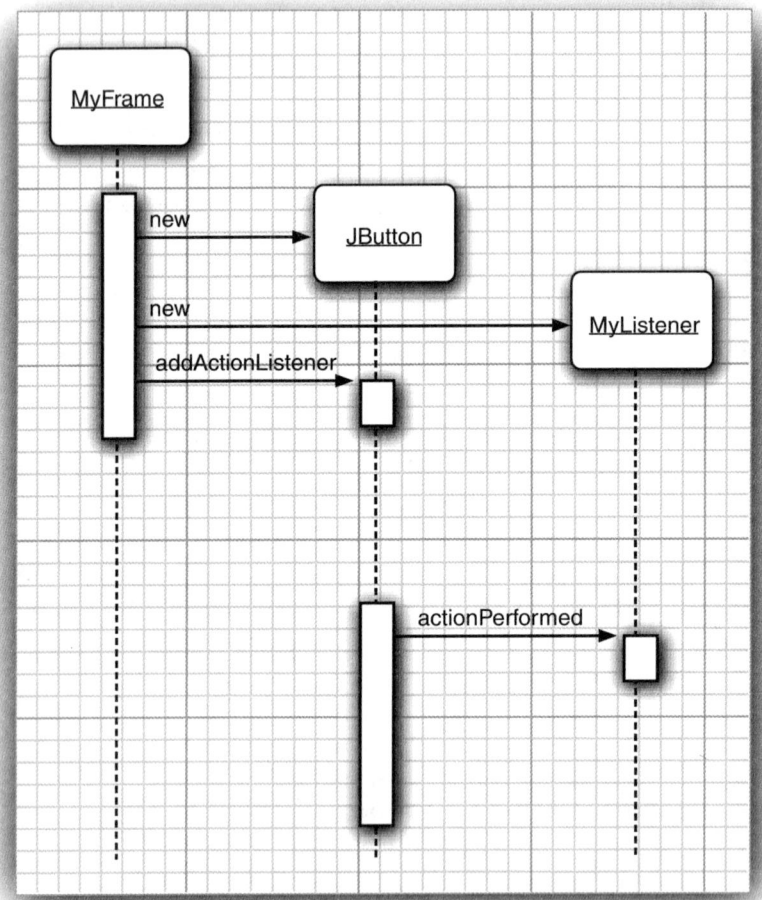

Figure 10.13 Event notification

10.4.2 Example: Handling a Button Click

As a way of getting comfortable with the event delegation model, let's work through all the details needed for the simple example of responding to a button click. For this example, we will show a panel populated with three buttons. Three listener objects are added as action listeners to the buttons.

With this scenario, each time a user clicks on any of the buttons on the panel, the associated listener object receives an ActionEvent that indicates a button click. In our sample program, the listener object will then change the background color of the panel.

Before we can show you the program that listens to button clicks, we first need to explain how to create buttons and how to add them to a panel.

To create a button, specify a label string, an icon, or both in the button constructor. Here are two examples:

```
var yellowButton = new JButton("Yellow");
var blueButton = new JButton(new ImageIcon("blue-ball.gif"));
```

Call the add method to add the buttons to a panel:

```
var yellowButton = new JButton("Yellow");
var blueButton = new JButton("Blue");
var redButton = new JButton("Red");

buttonPanel.add(yellowButton);
buttonPanel.add(blueButton);
buttonPanel.add(redButton);
```

Figure 10.14 shows the result.

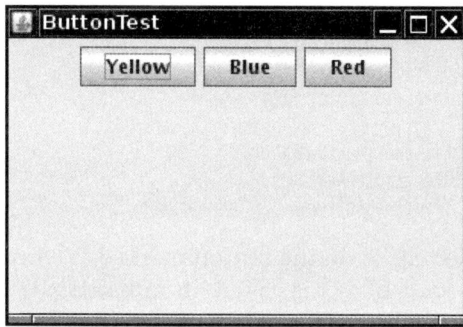

Figure 10.14 A panel filled with buttons

Next, we need to add code that listens to these buttons. This requires classes that implement the ActionListener interface, which, as we just mentioned, has one method: actionPerformed, whose signature looks like this:

```
public void actionPerformed(ActionEvent event)
```

The way to use the ActionListener interface is the same in all situations: The actionPerformed method (which is the only method in ActionListener) takes an object of type ActionEvent as a parameter. This event object gives you information about the event that happened.

When a button is clicked, we want the background color of the panel to change to a particular color. We store the desired color in our listener class.

```
class ColorAction implements ActionListener
{
    private Color backgroundColor;

    public ColorAction(Color c)
    {
        backgroundColor = c;
    }

    public void actionPerformed(ActionEvent event)
    {
        // set panel background color
        . . .
    }
}
```

We then construct one object for each color and set the objects as the button listeners.

```
var yellowAction = new ColorAction(Color.YELLOW);
var blueAction = new ColorAction(Color.BLUE);
var redAction = new ColorAction(Color.RED);

yellowButton.addActionListener(yellowAction);
blueButton.addActionListener(blueAction);
redButton.addActionListener(redAction);
```

For example, if a user clicks on the button marked "Yellow", the actionPerformed method of the yellowAction object is called. Its backgroundColor instance field is set to Color.YELLOW, and it can now proceed to set the panel's background color.

Just one issue remains. The ColorAction object doesn't have access to the buttonPanel variable. You can solve this problem in two ways. You can store the panel in the ColorAction object and set it in the ColorAction constructor. Or, more conveniently, you can make ColorAction into an inner class of the ButtonFrame class. Its methods can then access the outer panel automatically.

Listing 10.5 contains the complete frame class. Whenever you click one of the buttons, the appropriate action listener changes the background color of the panel.

Listing 10.5 button/ButtonFrame.java

```java
1  package button;
2
3  import java.awt.*;
4  import java.awt.event.*;
5  import javax.swing.*;
6
7  /**
8   * A frame with a button panel.
9   */
10 public class ButtonFrame extends JFrame
11 {
12    private JPanel buttonPanel;
13    private static final int DEFAULT_WIDTH = 300;
14    private static final int DEFAULT_HEIGHT = 200;
15
16    public ButtonFrame()
17    {
18       setSize(DEFAULT_WIDTH, DEFAULT_HEIGHT);
19
20       // create buttons
21       var yellowButton = new JButton("Yellow");
22       var blueButton = new JButton("Blue");
23       var redButton = new JButton("Red");
24
25       buttonPanel = new JPanel();
26
27       // add buttons to panel
28       buttonPanel.add(yellowButton);
29       buttonPanel.add(blueButton);
30       buttonPanel.add(redButton);
31
32       // add panel to frame
33       add(buttonPanel);
34
35       // create button actions
36       var yellowAction = new ColorAction(Color.YELLOW);
37       var blueAction = new ColorAction(Color.BLUE);
38       var redAction = new ColorAction(Color.RED);
39
40       // associate actions with buttons
41       yellowButton.addActionListener(yellowAction);
42       blueButton.addActionListener(blueAction);
43       redButton.addActionListener(redAction);
44    }
45
```

(Continues)

Listing 10.5 *(Continued)*

```
46    /**
47     * An action listener that sets the panel's background color.
48     */
49    private class ColorAction implements ActionListener
50    {
51       private Color backgroundColor;
52
53       public ColorAction(Color c)
54       {
55          backgroundColor = c;
56       }
57
58       public void actionPerformed(ActionEvent event)
59       {
60          buttonPanel.setBackground(backgroundColor);
61       }
62    }
63 }
```

---

javax.swing.JButton 1.2

- JButton(String label)
- JButton(Icon icon)
- JButton(String label, Icon icon)

    constructs a button. The label string can be plain text or HTML; for example, "\<html>\Ok\\</html>".

---

java.awt.Container 1.0

- Component add(Component c)

    adds the component c to this container.

10.4.3 Specifying Listeners Concisely

In the preceding section, we defined a class for the event listener and constructed three objects of that class. It is not all that common to have multiple instances of a listener class. Most commonly, each listener carries out a separate action. In that case, there is no need to make a separate class. Simply use a lambda expression:

```
exitButton.addActionListener(event -> System.exit(0));
```

Now consider the case in which we have multiple related actions, such as the color buttons of the preceding section. In such a case, implement a helper method:

```
public void makeButton(String name, Color backgroundColor)
{
   var button = new JButton(name);
   buttonPanel.add(button);
   button.addActionListener(event ->
      buttonPanel.setBackground(backgroundColor));
}
```

Note that the lambda expression refers to the parameter variable backgroundColor.

Then we simply call

```
makeButton("yellow", Color.YELLOW);
makeButton("blue", Color.BLUE);
makeButton("red", Color.RED);
```

Here, we construct three listener objects, one for each color, without explicitly defining a class. Each time the helper method is called, it makes an instance of a class that implements the ActionListener interface. Its actionPerformed action references the backGroundColor value that is, in fact, stored with the listener object. However, all this happens without you having to explicitly define listener classes, instance variables, or constructors that set them.

 NOTE: In older code, you will often see the use of anonymous classes:

```
exitButton.addActionListener(new ActionListener()
   {
      public void actionPerformed(new ActionEvent)
      {
         System.exit(0);
      }
   });
```

Of course, this rather verbose code is no longer necessary. Using a lambda expression is simpler and clearer.

10.4.4 Adapter Classes

Not all events are as simple to handle as button clicks. Suppose you want to monitor when the user tries to close the main frame in order to put up a dialog and exit the program only when the user agrees.

When the user tries to close a window, the JFrame object is the source of a WindowEvent. If you want to catch that event, you must have an appropriate listener object and add it to the frame's list of window listeners.

```
WindowListener listener = . . .;
frame.addWindowListener(listener);
```

The window listener must be an object of a class that implements the WindowListener interface. There are actually seven methods in the WindowListener interface. The frame calls them as the responses to seven distinct events that could happen to a window. The names are self-explanatory, except that "iconified" is usually called "minimized" under Windows. Here is the complete WindowListener interface:

```
public interface WindowListener
{
   void windowOpened(WindowEvent e);
   void windowClosing(WindowEvent e);
   void windowClosed(WindowEvent e);
   void windowIconified(WindowEvent e);
   void windowDeiconified(WindowEvent e);
   void windowActivated(WindowEvent e);
   void windowDeactivated(WindowEvent e);
}
```

Of course, we can define a class that implements the interface, add a call to System.exit(0) in the windowClosing method, and write do-nothing functions for the other six methods. However, typing code for six methods that don't do anything is the kind of tedious busywork that nobody likes. To simplify this task, each of the AWT listener interfaces that have more than one method comes with a companion *adapter* class that implements all the methods in the interface but does nothing with them. For example, the WindowAdapter class has seven do-nothing methods. You extend the adapter class to specify the desired reactions to some, but not all, of the event types in the interface. (An interface such as ActionListener that has only a single method does not need an adapter class.)

Here is how we can define a window listener that overrides the windowClosing method:

```
class Terminator extends WindowAdapter
{
   public void windowClosing(WindowEvent e)
   {
      if (user agrees)
         System.exit(0);
   }
}
```

Now you can register an object of type Terminator as the event listener:

```
var listener = new Terminator();
frame.addWindowListener(listener);
```

 NOTE: Nowadays, one would implement do-nothing methods of the WindowListener interface as default methods. However, Swing was invented many years before there were default methods.

---

java.awt.event.WindowListener 1.1

- void windowOpened(WindowEvent e)

  is called after the window has been opened.

- void windowClosing(WindowEvent e)

  is called when the user has issued a window manager command to close the window. Note that the window will close only if its hide or dispose method is called.

- void windowClosed(WindowEvent e)

  is called after the window has closed.

- void windowIconified(WindowEvent e)

  is called after the window has been iconified.

- void windowDeiconified(WindowEvent e)

  is called after the window has been deiconified.

- void windowActivated(WindowEvent e)

  is called after the window has become active. Only a frame or dialog can be active. Typically, the window manager decorates the active window—for example, by highlighting the title bar.

- void windowDeactivated(WindowEvent e)

  is called after the window has become deactivated.

---

java.awt.event.WindowStateListener 1.4

- void windowStateChanged(WindowEvent event)

  is called after the window has been maximized, iconified, or restored to normal size.

10.4.5 Actions

It is common to have multiple ways to activate the same command. The user can choose a certain function through a menu, a keystroke, or a button on a toolbar. This is easy to achieve in the AWT event model: link all events to the same listener. For example, suppose blueAction is an action listener whose actionPerformed method changes the background color to blue. You can attach the same object as a listener to several event sources:

- A toolbar button labeled "Blue"
- A menu item labeled "Blue"
- A keystroke Ctrl+B

The color change command will now be handled in a uniform way, no matter whether it was caused by a button click, a menu selection, or a key press.

The Swing package provides a very useful mechanism to encapsulate commands and to attach them to multiple event sources: the Action interface. An *action* is an object that encapsulates

- A description of the command (as a text string and an optional icon); and
- Parameters that are necessary to carry out the command (such as the requested color in our example).

The Action interface has the following methods:

```
void actionPerformed(ActionEvent event)
void setEnabled(boolean b)
boolean isEnabled()
void putValue(String key, Object value)
Object getValue(String key)
void addPropertyChangeListener(PropertyChangeListener listener)
void removePropertyChangeListener(PropertyChangeListener listener)
```

The first method is the familiar method in the ActionListener interface; in fact, the Action interface extends the ActionListener interface. Therefore, you can use an Action object whenever an ActionListener object is expected.

The next two methods let you enable or disable the action and check whether the action is currently enabled. When an action is attached to a menu or toolbar and the action is disabled, the option is grayed out.

The putValue and getValue methods let you store and retrieve arbitrary name/value pairs in the action object. A couple of important predefined strings, namely Action.NAME and Action.SMALL_ICON, store action names and icons into an action object:

```
action.putValue(Action.NAME, "Blue");
action.putValue(Action.SMALL_ICON, new ImageIcon("blue-ball.gif"));
```

Table 10.1 shows all predefined action table names.

Table 10.1 Predefined Action Table Names

Name	Value
NAME	The name of the action, displayed on buttons and menu items.
SMALL_ICON	A place to store a small icon for display in a button, menu item, or toolbar.
SHORT_DESCRIPTION	A short description of the icon for display in a tooltip.
LONG_DESCRIPTION	A long description of the icon for potential use in online help. No Swing component uses this value.
MNEMONIC_KEY	A mnemonic abbreviation for display in menu items.
ACCELERATOR_KEY	A place to store an accelerator keystroke. No Swing component uses this value.
ACTION_COMMAND_KEY	Historically, used in the now-obsolete registerKeyboardAction method.
DEFAULT	Potentially useful catch-all property. No Swing component uses this value.

If the action object is added to a menu or toolbar, the name and icon are automatically retrieved and displayed in the menu item or toolbar button. The SHORT_DESCRIPTION value turns into a tooltip.

The final two methods of the Action interface allow other objects, in particular menus or toolbars that trigger the action, to be notified when the properties of the action object change. For example, if a menu is added as a property change listener of an action object and the action object is subsequently disabled, the menu is called and can gray out the action name.

Note that Action is an *interface*, not a class. Any class implementing this interface must implement the seven methods we just discussed. Fortunately, a friendly soul has provided a class AbstractAction that implements all methods except for actionPerformed. That class takes care of storing all name/value pairs and managing the property change listeners. You simply extend AbstractAction and supply an actionPerformed method.

Let's build an action object that can execute color change commands. We store the name of the command, an icon, and the desired color. We store the color in the table of name/value pairs that the AbstractAction class provides. Here is the code for the ColorAction class. The constructor sets the name/value pairs, and the actionPerformed method carries out the color change action.

```
public class ColorAction extends AbstractAction
{
   public ColorAction(String name, Icon icon, Color c)
   {
      putValue(Action.NAME, name);
      putValue(Action.SMALL_ICON, icon);
      putValue("color", c);
      putValue(Action.SHORT_DESCRIPTION, "Set panel color to " + name.toLowerCase());
   }

   public void actionPerformed(ActionEvent event)
   {
      Color c = (Color) getValue("color");
      buttonPanel.setBackground(c);
   }
}
```

Our test program creates three objects of this class, such as

```
var blueAction = new ColorAction("Blue", new ImageIcon("blue-ball.gif"), Color.BLUE);
```

Next, let's associate this action with a button. That is easy because we can use a JButton constructor that takes an Action object.

```
var blueButton = new JButton(blueAction);
```

That constructor reads the name and icon from the action, sets the short description as the tooltip, and sets the action as the listener. You can see the icons and a tooltip in Figure 10.15.

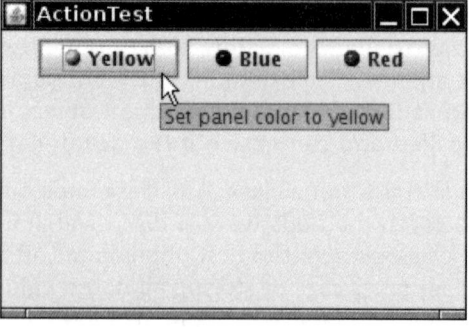

Figure 10.15 Buttons display the icons from the action objects.

As we demonstrate in the next chapter, it is just as easy to add the same action to a menu.

Finally, we want to add the action objects to keystrokes so that an action is carried out when the user types a keyboard command. To associate actions with keystrokes, you first need to generate objects of the KeyStroke class. This convenience class encapsulates the description of a key. To generate a KeyStroke object, don't call a constructor but instead use the static getKeyStroke method of the KeyStroke class.

```
KeyStroke ctrlBKey = KeyStroke.getKeyStroke("ctrl B");
```

To understand the next step, you need to understand the concept of *keyboard focus*. A user interface can have many buttons, menus, scrollbars, and other components. When you hit a key, it is sent to the component that has focus. That component is usually (but not always) visually distinguished. For example, in the Java look-and-feel, a button with focus has a thin rectangular border around the button text. You can use the Tab key to move the focus between components. When you press the space bar, the button with focus is clicked. Other keys carry out different actions; for example, the arrow keys can move a scrollbar.

However, in our case, we do not want to send the keystroke to the component that has focus. Otherwise, each of the buttons would need to know how to handle the Ctrl+Y, Ctrl+B, and Ctrl+R keys.

This is a common problem, and the Swing designers came up with a convenient solution. Every JComponent has three *input maps*, each mapping KeyStroke objects to associated actions. The three input maps correspond to three different conditions (see Table 10.2).

Table 10.2 Input Map Conditions

Flag	Invoke Action
WHEN_FOCUSED	When this component has keyboard focus
WHEN_ANCESTOR_OF_FOCUSED_COMPONENT	When this component contains the component that has keyboard focus
WHEN_IN_FOCUSED_WINDOW	When this component is contained in the same window as the component that has keyboard focus

Keystroke processing checks these maps in the following order:

1. Check the WHEN_FOCUSED map of the component with input focus. If the keystroke exists and its corresponding action is enabled, execute the action and stop processing.

2. Starting from the component with input focus, check the WHEN_ANCESTOR_OF_FOCUSED_COMPONENT maps of its parent components. As soon as a map with the keystroke and a corresponding enabled action is found, execute the action and stop processing.

3. Look at all *visible* and *enabled* components, in the window with input focus, that have this keystroke registered in a WHEN_IN_FOCUSED_WINDOW map. Give these components (in the order of their keystroke registration) a chance to execute the corresponding action. As soon as the first enabled action is executed, stop processing.

To obtain an input map from the component, use the getInputMap method. Here is an example:

```
InputMap imap = panel.getInputMap(JComponent.WHEN_FOCUSED);
```

The WHEN_FOCUSED condition means that this map is consulted when the current component has the keyboard focus. In our situation, that isn't the map we want. One of the buttons, not the panel, has the input focus. Either of the other two map choices works fine for inserting the color change keystrokes. We use WHEN_ANCESTOR_OF_FOCUSED_COMPONENT in our example program.

The InputMap doesn't directly map KeyStroke objects to Action objects. Instead, it maps to arbitrary objects, and a second map, implemented by the ActionMap class, maps objects to actions. That makes it easier to share the same actions among keystrokes that come from different input maps.

Thus, each component has three input maps and one action map. To tie them together, you need to come up with names for the actions. Here is how you can tie a key to an action:

```
imap.put(KeyStroke.getKeyStroke("ctrl Y"), "panel.yellow");
ActionMap amap = panel.getActionMap();
amap.put("panel.yellow", yellowAction);
```

It is customary to use the string "none" for a do-nothing action. That makes it easy to deactivate a key:

```
imap.put(KeyStroke.getKeyStroke("ctrl C"), "none");
```

CAUTION: The JDK documentation suggests using the action name as the action's key. We don't think that is a good idea. The action name is displayed on buttons and menu items; thus, it can change at the whim of the UI designer and may be translated into multiple languages. Such unstable strings are poor choices for lookup keys, so we recommend that you come up with action names that are independent of the displayed names.

To summarize, here is what you do to carry out the same action in response to a button, a menu item, or a keystroke:

1. Implement a class that extends the AbstractAction class. You may be able to use the same class for multiple related actions.
2. Construct an object of the action class.
3. Construct a button or menu item from the action object. The constructor will read the label text and icon from the action object.
4. For actions that can be triggered by keystrokes, you have to carry out additional steps. First, locate the top-level component of the window, such as a panel that contains all other components.
5. Then, get the WHEN_ANCESTOR_OF_FOCUSED_COMPONENT input map of the top-level component. Make a KeyStroke object for the desired keystroke. Make an action key object, such as a string that describes your action. Add the pair (keystroke, action key) into the input map.
6. Finally, get the action map of the top-level component. Add the pair (action key, action object) into the map.

---

javax.swing.Action 1.2

- boolean isEnabled()
- void setEnabled(boolean b)

  gets or sets the enabled property of this action.

- void putValue(String key, Object value)

  places a key/value pair inside the action object. The key can be any string, but several names have predefined meanings—see Table 10.1.

- Object getValue(String key)

  returns the value of a stored name/value pair.

`javax.swing.KeyStroke` 1.2

- `static KeyStroke getKeyStroke(String description)`

    constructs a keystroke from a human-readable description (a sequence of whitespace-delimited strings). The description starts with zero or more modifiers (shift, control, ctrl, meta, alt, altGraph) and ends with either the string typed, followed by a one-character string (for example, "typed a"), or an optional event specifier (pressed or released, with pressed being the default), followed by a key code. The key code, when prefixed with VK_, should correspond to a KeyEvent constant; for example, "INSERT" corresponds to KeyEvent.VK_INSERT.

`javax.swing.JComponent` 1.2

- `ActionMap getActionMap()` 1.3

    returns the map that associates action map keys (which can be arbitrary objects) with Action objects.

- `InputMap getInputMap(int flag)` 1.3

    gets the input map that maps key strokes to action map keys. The flag is one of the values in Table 10.2.

10.4.6 Mouse Events

You do not need to handle mouse events explicitly if you just want the user to be able to click on a button or menu. These mouse operations are handled internally by the various components in the user interface. However, if you want to enable the user to draw with the mouse, you will need to trap the mouse move, click, and drag events.

In this section, we will show you a simple graphics editor application that allows the user to place, move, and erase squares on a canvas (see Figure 10.16).

When the user clicks a mouse button, three listener methods are called: mousePressed when the mouse is first pressed, mouseReleased when the mouse is released, and, finally, mouseClicked. If you are only interested in complete clicks, you can ignore the first two methods. By using the getX and getY methods on the MouseEvent argument, you can obtain the x and y coordinates of the mouse pointer when the mouse was clicked. To distinguish between single, double, and triple (!) clicks, use the getClickCount method.

In our sample program, we supply both a mousePressed and a mouseClicked methods. When you click on a pixel that is not inside any of the squares that have

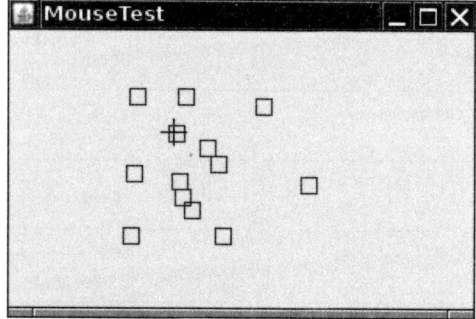

Figure 10.16 A mouse test program

been drawn, a new square is added. We implemented this in the mousePressed method so that the user receives immediate feedback and does not have to wait until the mouse button is released. When a user double-clicks inside an existing square, it is erased. We implemented this in the mouseClicked method because we need the click count.

```
public void mousePressed(MouseEvent event)
{
   current = find(event.getPoint());
   if (current == null) // not inside a square
      add(event.getPoint());
}

public void mouseClicked(MouseEvent event)
{
   current = find(event.getPoint());
   if (current != null && event.getClickCount() >= 2)
      remove(current);
}
```

As the mouse moves over a window, the window receives a steady stream of mouse movement events. Note that there are separate MouseListener and MouseMotionListener interfaces. This is done for efficiency—there are a lot of mouse events as the user moves the mouse around, and a listener that just cares about mouse *clicks* will not be bothered with unwanted mouse *moves*.

Our test application traps mouse motion events to change the cursor to a different shape (a cross hair) when it is over a square. This is done with the getPredefinedCursor method of the Cursor class. Table 10.3 lists the constants to use with this method along with what the cursors look like under Windows.

Here is the mouseMoved method of the MouseMotionListener in our example program:

Table 10.3 Sample Cursor Shapes

Icon	Constant	Icon	Constant
	DEFAULT_CURSOR		NE_RESIZE_CURSOR
	CROSSHAIR_CURSOR		E_RESIZE_CURSOR
	HAND_CURSOR		SE_RESIZE_CURSOR
	MOVE_CURSOR		S_RESIZE_CURSOR
	TEXT_CURSOR		SW_RESIZE_CURSOR
	WAIT_CURSOR		W_RESIZE_CURSOR
	N_RESIZE_CURSOR		NW_RESIZE_CURSOR

```
public void mouseMoved(MouseEvent event)
{
   if (find(event.getPoint()) == null)
      setCursor(Cursor.getDefaultCursor());
   else
      setCursor(Cursor.getPredefinedCursor(Cursor.CROSSHAIR_CURSOR));
}
```

If the user presses a mouse button while the mouse is in motion, mouseDragged calls are generated instead of mouseMoved calls. Our test application lets a user drag the square under the cursor. We simply update the currently dragged rectangle to be centered under the mouse position. Then, we repaint the canvas to show the new mouse position.

```
public void mouseDragged(MouseEvent event)
{
   if (current != null)
   {
      int x = event.getX();
      int y = event.getY();

      current.setFrame(x - SIDELENGTH / 2, y - SIDELENGTH / 2, SIDELENGTH, SIDELENGTH);
      repaint();
   }
}
```

NOTE: The mouseMoved method is only called as long as the mouse stays inside the component. However, the mouseDragged method keeps getting called even when the mouse is being dragged outside the component.

There are two other mouse event methods: mouseEntered and mouseExited. These methods are called when the mouse enters or exits a component.

Finally, we explain how to listen to mouse events. Mouse clicks are reported through the mouseClicked method, which is part of the MouseListener interface. Many applications are only interested in mouse clicks and not in mouse moves; with the mouse move events occurring so frequently, the mouse move and drag events are defined in a separate interface called MouseMotionListener.

In our program we are interested in both types of mouse events. We define two inner classes: MouseHandler and MouseMotionHandler. The MouseHandler class extends the MouseAdapter class because it defines only two of the five MouseListener methods. The MouseMotionHandler implements the MouseMotionListener and defines both methods of that interface. Listing 10.6 is the program listing.

Listing 10.6 mouse/MouseComponent.java

```
 1  package mouse;
 2
 3  import java.awt.*;
 4  import java.awt.event.*;
 5  import java.awt.geom.*;
 6  import java.util.*;
 7  import javax.swing.*;
 8
 9  /**
10   * A component with mouse operations for adding and removing squares.
11   */
12  public class MouseComponent extends JComponent
13  {
14     private static final int DEFAULT_WIDTH = 300;
15     private static final int DEFAULT_HEIGHT = 200;
16
17     private static final int SIDELENGTH = 10;
18     private ArrayList<Rectangle2D> squares;
19     private Rectangle2D current; // the square containing the mouse cursor
20
21     public MouseComponent()
22     {
23        squares = new ArrayList<>();
24        current = null;
```

(Continues)

Listing 10.6 *(Continued)*

```
25
26        addMouseListener(new MouseHandler());
27        addMouseMotionListener(new MouseMotionHandler());
28     }
29
30     public Dimension getPreferredSize()
31     {
32        return new Dimension(DEFAULT_WIDTH, DEFAULT_HEIGHT);
33     }
34
35     public void paintComponent(Graphics g)
36     {
37        var g2 = (Graphics2D) g;
38
39        // draw all squares
40        for (Rectangle2D r : squares)
41           g2.draw(r);
42     }
43
44     /**
45      * Finds the first square containing a point.
46      * @param p a point
47      * @return the first square that contains p
48      */
49     public Rectangle2D find(Point2D p)
50     {
51        for (Rectangle2D r : squares)
52        {
53           if (r.contains(p)) return r;
54        }
55        return null;
56     }
57
58     /**
59      * Adds a square to the collection.
60      * @param p the center of the square
61      */
62     public void add(Point2D p)
63     {
64        double x = p.getX();
65        double y = p.getY();
66
67        current = new Rectangle2D.Double(x - SIDELENGTH / 2, y - SIDELENGTH / 2,
68           SIDELENGTH, SIDELENGTH);
69        squares.add(current);
70        repaint();
71     }
```

```
72
73    /**
74     * Removes a square from the collection.
75     * @param s the square to remove
76     */
77    public void remove(Rectangle2D s)
78    {
79       if (s == null) return;
80       if (s == current) current = null;
81       squares.remove(s);
82       repaint();
83    }
84
85    private class MouseHandler extends MouseAdapter
86    {
87       public void mousePressed(MouseEvent event)
88       {
89          // add a new square if the cursor isn't inside a square
90          current = find(event.getPoint());
91          if (current == null) add(event.getPoint());
92       }
93
94       public void mouseClicked(MouseEvent event)
95       {
96          // remove the current square if double clicked
97          current = find(event.getPoint());
98          if (current != null && event.getClickCount() >= 2) remove(current);
99       }
100   }
101
102   private class MouseMotionHandler implements MouseMotionListener
103   {
104      public void mouseMoved(MouseEvent event)
105      {
106         // set the mouse cursor to cross hairs if it is inside a rectangle
107
108         if (find(event.getPoint()) == null) setCursor(Cursor.getDefaultCursor());
109         else setCursor(Cursor.getPredefinedCursor(Cursor.CROSSHAIR_CURSOR));
110      }
111
112      public void mouseDragged(MouseEvent event)
113      {
114         if (current != null)
115         {
116            int x = event.getX();
117            int y = event.getY();
118
119            // drag the current rectangle to center it at (x, y)
```

(Continues)

Listing 10.6 *(Continued)*

```
120              current.setFrame(x - SIDELENGTH / 2, y - SIDELENGTH / 2, SIDELENGTH, SIDELENGTH);
121              repaint();
122           }
123        }
124     }
125 }
```

---

java.awt.event.MouseEvent 1.1

- int getX()
- int getY()
- Point getPoint()

  returns the x (horizontal) and y (vertical) coordinates of the point where the event happened, measured from the top left corner of the component that is the event source.

- int getClickCount()

  returns the number of consecutive mouse clicks associated with this event. (The time interval for what constitutes "consecutive" is system-dependent.)

---

java.awt.Component 1.0

- public void setCursor(Cursor cursor) 1.1

  sets the cursor image to the specified cursor.

10.4.7 The AWT Event Hierarchy

The EventObject class has a subclass AWTEvent, which is the parent of all AWT event classes. Figure 10.17 shows the inheritance diagram of the AWT events.

Some of the Swing components generate event objects of yet more event types; these directly extend EventObject, not AWTEvent.

The event objects encapsulate information about the event that the event source communicates to its listeners. When necessary, you can then analyze the event objects that were passed to the listener object, as we did in the button example with the getSource and getActionCommand methods.

Some of the AWT event classes are of no practical use for the Java programmer. For example, the AWT inserts PaintEvent objects into the event queue, but these objects are not delivered to listeners. Java programmers don't listen

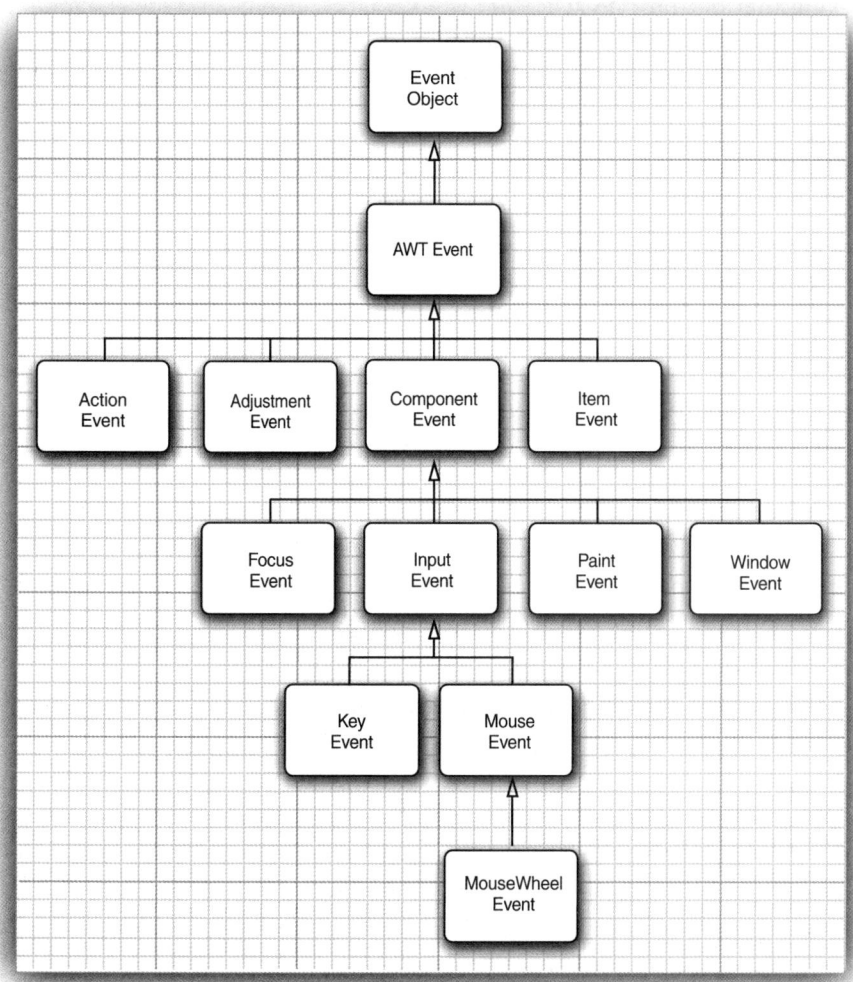

Figure 10.17 Inheritance diagram of AWT event classes

to paint events; instead, they override the `paintComponent` method to control repainting. The AWT also generates a number of events that are needed only by systems programmers, to provide input systems for ideographic languages, automated testing robots, and so on.

The AWT makes a useful distinction between *low-level* and *semantic* events. A semantic event is one that expresses what the user is doing, such as "clicking that button"; an ActionEvent is a semantic event. Low-level events are those events that make this possible. In the case of a button click, this is a mouse down, a series of mouse moves, and a mouse up (but only if the mouse up is inside the button area). Or it might be a keystroke, which happens if the user selects the button with the Tab key and then activates it with the space bar. Similarly, adjusting a scrollbar is a semantic event, but dragging the mouse is a low-level event.

Here are the most commonly used semantic event classes in the java.awt.event package:

- ActionEvent (for a button click, a menu selection, selecting a list item, or Enter typed in a text field)
- AdjustmentEvent (the user adjusted a scrollbar)
- ItemEvent (the user made a selection from a set of checkbox or list items)

Five low-level event classes are commonly used:

- KeyEvent (a key was pressed or released)
- MouseEvent (the mouse button was pressed, released, moved, or dragged)
- MouseWheelEvent (the mouse wheel was rotated)
- FocusEvent (a component got focus or lost focus)
- WindowEvent (the window state changed)

Table 10.4 shows the most important AWT listener interfaces, events, and event sources.

Table 10.4 Event Handling Summary

Interface	Methods	Parameter/Accessors	Events Generated By
ActionListener	actionPerformed	ActionEvent • getActionCommand • getModifiers	AbstractButton JComboBox JTextField Timer
AdjustmentListener	adjustmentValueChanged	AdjustmentEvent • getAdjustable • getAdjustmentType • getValue	JScrollbar

(Continues)

Table 10.4 *(Continued)*

Interface	Methods	Parameter/Accessors	Events Generated By
ItemListener	itemStateChanged	ItemEvent • getItem • getItemSelectable • getStateChange	AbstractButton JComboBox
FocusListener	focusGained focusLost	FocusEvent • isTemporary	Component
KeyListener	keyPressed keyReleased keyTyped	KeyEvent • getKeyChar • getKeyCode • getKeyModifiersText • getKeyText • isActionKey	Component
MouseListener	mousePressed mouseReleased mouseEntered mouseExited mouseClicked	MouseEvent • getClickCount • getX • getY • getPoint • translatePoint	Component
MouseMotionListener	mouseDragged mouseMoved	MouseEvent	Component
MouseWheelListener	mouseWheelMoved	MouseWheelEvent • getWheelRotation • getScrollAmount	Component
WindowListener	windowClosing windowOpened windowIconified windowDeiconified windowClosed windowActivated windowDeactivated	WindowEvent • getWindow	Window

(Continues)

Table 10.4 *(Continued)*

Interface	Methods	Parameter/Accessors	Events Generated By
WindowFocusListener	windowGainedFocus windowLostFocus	WindowEvent • getOppositeWindow	Window
WindowStateListener	windowStateChanged	WindowEvent • getOldState • getNewState	Window

10.5 The Preferences API

We end this chapter with a discussion of the java.util.preferences API. In a desktop program, you will often want to store user preferences, such as the last file that the user worked on, the last window location, and so on.

As you have seen in Chapter 9, the Properties class makes it simple to load and save configuration information of a prorgram. However, using property files has these disadvantages:

• Some operating systems have no concept of a home directory, making it difficult to find a uniform location for configuration files.

• There is no standard convention for naming configuration files, increasing the likelihood of name clashes as users install multiple Java applications.

Some operating systems have a central repository for configuration information. The best-known example is the registry in Microsoft Windows. The Preferences class provides such a central repository in a platform-independent manner. In Windows, the Preferences class uses the registry for storage; on Linux, the information is stored in the local file system instead. Of course, the repository implementation is transparent to the programmer using the Preferences class.

The Preferences repository has a tree structure, with node path names such as /com/mycompany/myapp. As with package names, name clashes are avoided as long as programmers start the paths with reversed domain names. In fact, the designers of the API suggest that the configuration node paths match the package names in your program.

Each node in the repository has a separate table of key/value pairs that you can use to store numbers, strings, or byte arrays. No provision is made for storing serializable objects. The API designers felt that the serialization format is too fragile for long-term storage. Of course, if you disagree, you can save serialized objects in byte arrays.

For additional flexibility, there are multiple parallel trees. Each program user has one tree; an additional tree, called the system tree, is available for settings that are common to all users. The Preferences class uses the operating system's notion of the "current user" for accessing the appropriate user tree.

To access a node in the tree, start with the user or system root:

```
Preferences root = Preferences.userRoot();
```

or

```
Preferences root = Preferences.systemRoot();
```

Then access the node. You can simply provide a node path name:

```
Preferences node = root.node("/com/mycompany/myapp");
```

A convenient shortcut gets a node whose path name equals the package name of a class. Simply take an object of that class and call

```
Preferences node = Preferences.userNodeForPackage(obj.getClass());
```

or

```
Preferences node = Preferences.systemNodeForPackage(obj.getClass());
```

Typically, obj will be the this reference.

Once you have a node, you can access the key/value table with methods

```
String get(String key, String defval)
int getInt(String key, int defval)
long getLong(String key, long defval)
float getFloat(String key, float defval)
double getDouble(String key, double defval)
boolean getBoolean(String key, boolean defval)
byte[] getByteArray(String key, byte[] defval)
```

Note that you must specify a default value when reading the information, in case the repository data is not available. Defaults are required for several reasons. The data might be missing because the user never specified a preference. Certain resource-constrained platforms might not have a repository, and mobile devices might be temporarily disconnected from the repository.

Conversely, you can write data to the repository with put methods such as

```
put(String key, String value)
putInt(String key, int value)
```

and so on.

You can enumerate all keys stored in a node with the method

```
String[] keys()
```

There is currently no way to find out the type of the value of a particular key.

 NOTE: Node names and keys are limited to 80 characters, and string values to 8192 characters.

Central repositories such as the Windows registry traditionally suffer from two problems:

- They turn into a "dumping ground" filled with obsolete information.
- Configuration data gets entangled into the repository, making it difficult to move preferences to a new platform.

The Preferences class has a solution for the second problem. You can export the preferences of a subtree (or, less commonly, a single node) by calling the methods

```
void exportSubtree(OutputStream out)
void exportNode(OutputStream out)
```

The data are saved in XML format. You can import them into another repository by calling

```
void importPreferences(InputStream in)
```

Here is a sample file:

```
<?xml version="1.0" encoding="UTF-8"?>
<!DOCTYPE preferences SYSTEM "http://java.sun.com/dtd/preferences.dtd">
<preferences EXTERNAL_XML_VERSION="1.0">
  <root type="user">
    <map/>
    <node name="com">
      <map/>
      <node name="horstmann">
        <map/>
        <node name="corejava">
          <map>
          <entry key="height" value="200.0"/>
          <entry key="left" value="1027.0"/>
          <entry key="filename" value="/home/cay/books/cj11/code/v1ch11/raven.html"/>
          <entry key="top" value="380.0"/>
          <entry key="width" value="300.0"/>
            </map>
          </node>
```

```
            </node>
          </node>
        </root>
      </preferences>
```

If your program uses preferences, you should give your users the opportunity of exporting and importing them, so they can easily migrate their settings from one computer to another. The program in Listing 10.7 demonstrates this technique. The program simply saves the window location and the last loaded filename. Try resizing the window, then export your preferences, move the window, exit, and restart the application. The window will be just like you left it when you exited. Import your preferences, and the window reverts to its prior location.

Listing 10.7 preferences/ImageViewer.java

```java
1  package preferences;
2
3  import java.awt.EventQueue;
4  import java.awt.event.*;
5  import java.io.*;
6  import java.util.prefs.*;
7  import javax.swing.*;
8
9  /**
10  * A program to test preference settings. The program remembers the
11  * frame position, size, and last selected file.
12  * @version 1.10 2018-04-10
13  * @author Cay Horstmann
14  */
15  public class ImageViewer
16  {
17     public static void main(String[] args)
18     {
19        EventQueue.invokeLater(() -> {
20           var frame = new ImageViewerFrame();
21           frame.setTitle("ImageViewer");
22           frame.setDefaultCloseOperation(JFrame.EXIT_ON_CLOSE);
23           frame.setVisible(true);
24        });
25     }
26  }
27
28  /**
29  * An image viewer that restores position, size, and image from user
30  * preferences and updates the preferences upon exit.
31  */
```

(Continues)

Listing 10.7 (Continued)

```
32  class ImageViewerFrame extends JFrame
33  {
34     private static final int DEFAULT_WIDTH = 300;
35     private static final int DEFAULT_HEIGHT = 200;
36     private String image;
37
38     public ImageViewerFrame()
39     {
40        Preferences root = Preferences.userRoot();
41        Preferences node = root.node("/com/horstmann/corejava/ImageViewer");
42        // get position, size, title from properties
43        int left = node.getInt("left", 0);
44        int top = node.getInt("top", 0);
45        int width = node.getInt("width", DEFAULT_WIDTH);
46        int height = node.getInt("height", DEFAULT_HEIGHT);
47        setBounds(left, top, width, height);
48        image = node.get("image", null);
49        var label = new JLabel();
50        if (image != null) label.setIcon(new ImageIcon(image));
51
52        addWindowListener(new WindowAdapter()
53           {
54              public void windowClosing(WindowEvent event)
55              {
56                 node.putInt("left", getX());
57                 node.putInt("top", getY());
58                 node.putInt("width", getWidth());
59                 node.putInt("height", getHeight());
60                 node.put("image", image);
61              }
62           });
63
64        // use a label to display the images
65        add(label);
66
67        // set up the file chooser
68        var chooser = new JFileChooser();
69        chooser.setCurrentDirectory(new File("."));
70
71        // set up the menu bar
72        var menuBar = new JMenuBar();
73        setJMenuBar(menuBar);
74
75        var menu = new JMenu("File");
76        menuBar.add(menu);
```

```
77
78        var openItem = new JMenuItem("Open");
79        menu.add(openItem);
80        openItem.addActionListener(event -> {
81
82           // show file chooser dialog
83           int result = chooser.showOpenDialog(null);
84
85           // if file selected, set it as icon of the label
86           if (result == JFileChooser.APPROVE_OPTION)
87           {
88              image = chooser.getSelectedFile().getPath();
89              label.setIcon(new ImageIcon(image));
90           }
91        });
92
93        var exitItem = new JMenuItem("Exit");
94        menu.add(exitItem);
95        exitItem.addActionListener(event -> System.exit(0));
96     }
97  }
```

---

java.util.prefs.Preferences 1.4

- Preferences userRoot()

  returns the root preferences node of the user of the calling program.

- Preferences systemRoot()

  returns the systemwide root preferences node.

- Preferences node(String path)

  returns a node that can be reached from the current node by the given path. If path is absolute (that is, starts with a /), then the node is located starting from the root of the tree containing this preference node. If there isn't a node with the given path, it is created.

- Preferences userNodeForPackage(Class cl)
- Preferences systemNodeForPackage(Class cl)

  returns a node in the current user's tree or the system tree whose absolute node path corresponds to the package name of the class cl.

- String[] keys()

  returns all keys belonging to this node.

(Continues)

---

java.util.prefs.Preferences 1.4 *(Continued)*

- String get(String key, String defval)
- int getInt(String key, int defval)
- long getLong(String key, long defval)
- float getFloat(String key, float defval)
- double getDouble(String key, double defval)
- boolean getBoolean(String key, boolean defval)
- byte[] getByteArray(String key, byte[] defval)

    returns the value associated with the given key or the supplied default value if no value is associated with the key, the associated value is not of the correct type, or the preferences store is unavailable.

- void put(String key, String value)
- void putInt(String key, int value)
- void putLong(String key, long value)
- void putFloat(String key, float value)
- void putDouble(String key, double value)
- void putBoolean(String key, boolean value)
- void putByteArray(String key, byte[] value)

    stores a key/value pair with this node.

- void exportSubtree(OutputStream out)

    writes the preferences of this node and its children to the specified stream.

- void exportNode(OutputStream out)

    writes the preferences of this node (but not its children) to the specified stream.

- void importPreferences(InputStream in)

    imports the preferences contained in the specified stream.

This concludes our introduction into graphical user interface programming. The next chapter shows you how to work with the most common Swing components.

CHAPTER

User Interface Components with Swing

In this chapter

The previous chapter was written primarily to show you how to use the event model in Java. In the process, you took the first steps toward learning how to build a graphical user interface. This chapter shows you the most important tools you'll need to build more full-featured GUIs.

We start out with a tour of the architectural underpinnings of Swing. Knowing what goes on "under the hood" is important in understanding how to use some of the more advanced components effectively. We then show you the most common user interface components in Swing, such as text fields, radio buttons, and menus. Next, you will learn how to use layout managers to arrange these components. Finally, you'll see how to implement dialog boxes in Swing.

This chapter covers the basic Swing components such as text components, buttons, and sliders. These are the essential user interface components that you will need most frequently. We will cover advanced Swing components in Volume II.

11.1 Swing and the Model–View–Controller Design Pattern

Let's step back for a minute and think about the pieces that make up a user interface component such as a button, a checkbox, a text field, or a sophisticated tree control. Every component has three characteristics:

- Its *content*, such as the state of a button (pushed in or not), or the text in a text field
- Its *visual appearance* (color, size, and so on)
- Its *behavior* (reaction to events)

Even a seemingly simple component such as a button exhibits some moderately complex interaction among these characteristics. Obviously, the visual appearance of a button depends on the look-and-feel. A Metal button looks different from a Windows button or a Motif button. In addition, the appearance depends on the button state; when a button is pushed in, it needs to be redrawn to look different. The state depends on the events that the button receives. When the user depresses the mouse inside the button, the button is pushed in.

Of course, when you use a button in your programs, you simply consider it as a *button*; you don't think too much about the inner workings and characteristics. That, after all, is the job of the programmer who implemented the button. However, programmers who implement buttons and all other user interface components are motivated to think a little harder about them, so that they work well no matter what look-and-feel is in effect.

To do this, the Swing designers turned to a well-known design pattern: the *model-view-controller* (MVC) pattern. This design pattern tells us to provide three separate objects:

- The *model,* which stores the content
- The *view,* which displays the content
- The *controller,* which handles user input

The pattern specifies precisely how these three objects interact. The model stores the content and has *no user interface*. For a button, the content is pretty trivial—just a small set of flags that tells whether the button is currently pushed in or out, whether it is active or inactive, and so on. For a text field, the content is a bit more interesting. It is a string object that holds the current text. This is *not the same* as the view of the content—if the content is larger than the text field, the user sees only a portion of the text displayed (see Figure 11.1).

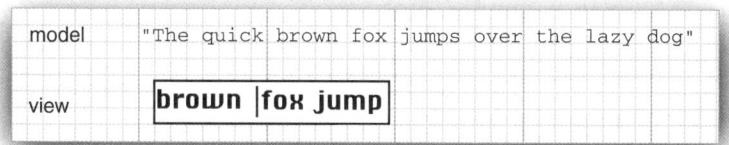

Figure 11.1 Model and view of a text field

The model must implement methods to change the content and to discover what the content is. For example, a text model has methods to add or remove characters in the current text and to return the current text as a string. Again, keep in mind that the model is completely nonvisual. It is the job of a view to draw the data stored in the model.

 NOTE: The term "model" is perhaps unfortunate because we often think of a model as a representation of an abstract concept. Car and airplane designers build models to simulate real cars and planes. But that analogy really leads you astray when thinking about the model-view-controller pattern. In this design pattern, the model stores the complete content, and the view gives a (complete or incomplete) visual representation of the content. A better analogy might be the model who poses for an artist. It is up to the artist to look at the model and create a view. Depending on the artist, that view might be a formal portrait, an impressionist painting, or a cubist drawing with strangely contorted limbs.

One of the advantages of the model-view-controller pattern is that a model can have multiple views, each showing a different part or aspect of the full content. For example, an HTML editor can offer two *simultaneous* views of the same content: a WYSIWYG view and a "raw tag" view (see Figure 11.2). When the model is updated through the controller of one of the views, it tells both attached views about the change. When the views are notified, they refresh themselves automatically. Of course, for a simple user interface component such as a button, you won't have multiple views of the same model.

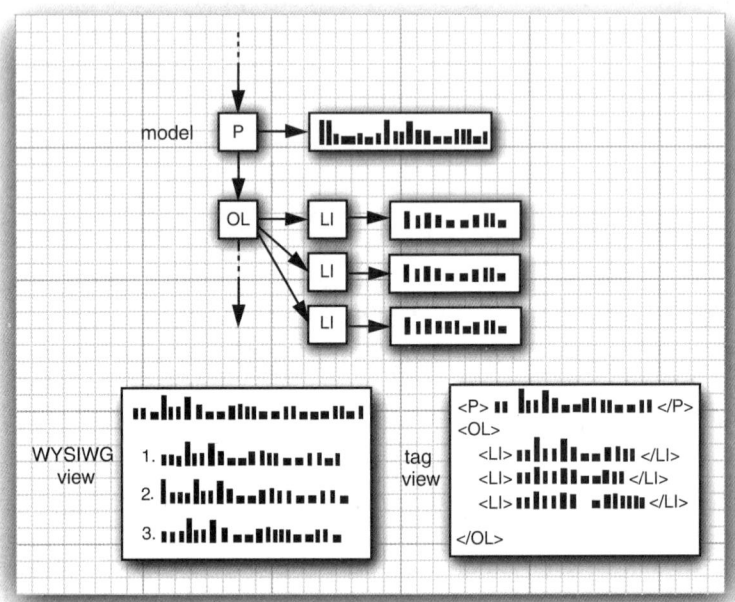

Figure 11.2 Two separate views of the same model

The controller handles the user-input events, such as mouse clicks and keystrokes. It then decides whether to translate these events into changes in the model or the view. For example, if the user presses a character key in a text box, the controller calls the "insert character" command of the model. The model then tells the view to update itself. The view never knows why the text changed. But if the user presses a cursor key, the controller may tell the view to scroll. Scrolling the view has no effect on the underlying text, so the model never knows that this event happened.

Figure 11.3 shows the interactions among model, view, and controller objects.

For most Swing components, the model class implements an interface whose name ends in Model; in this case, the interface is called ButtonModel. Classes implementing that interface can define the state of the various kinds of buttons. Actually, buttons aren't all that complicated, and the Swing library contains a single class, called DefaultButtonModel, that implements this interface.

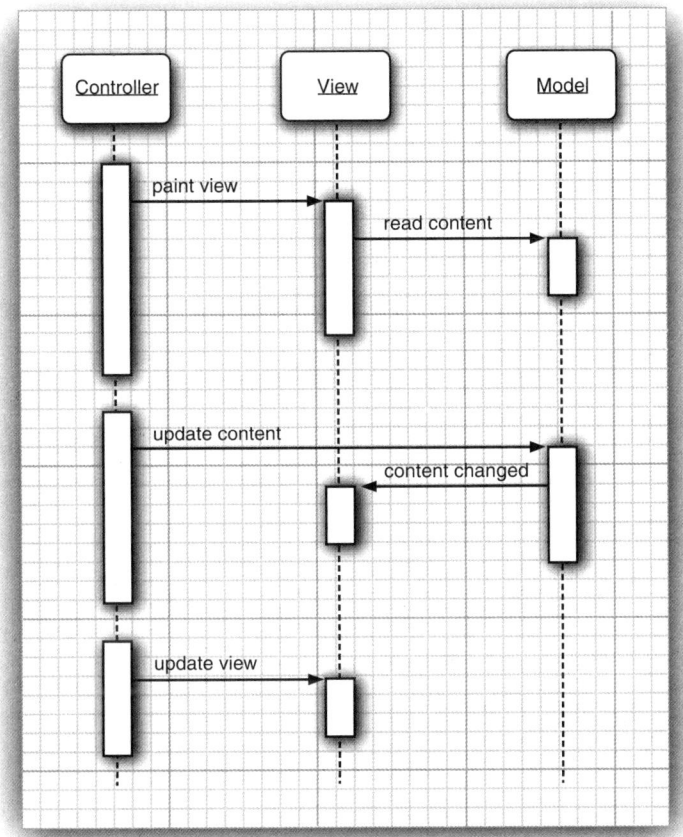

Figure 11.3 Interactions among model, view, and controller objects

You can get a sense of the sort of data maintained by a button model by looking at the properties of the ButtonModel interface—see Table 11.1.

Each JButton object stores a button model object which you can retrieve.

```
var button = new JButton("Blue");
ButtonModel model = button.getModel();
```

In practice, you won't care—the minutiae of the button state are only of interest to the view that draws it. All the important information—such as whether a button is enabled—is available from the JButton class. (Of course, the JButton then asks its model to retrieve that information.)

Table 11.1 Properties of the ButtonModel Interface

Property Name	Value
actionCommand	The action command string associated with this button
mnemonic	The keyboard mnemonic for this button
armed	true if the button was pressed and the mouse is still over the button
enabled	true if the button is selectable
pressed	true if the button was pressed but the mouse button hasn't yet been released
rollover	true if the mouse is over the button
selected	true if the button has been toggled on (used for checkboxes and radio buttons)

Have another look at the ButtonModel interface to see what *isn't* there. The model does *not* store the button label or icon. There is no way to find out what's on the face of a button just by looking at its model. (Actually, as you will see in Section 11.4.2, "Radio Buttons," on p. 654, this purity of design is the source of some grief for the programmer.)

It is also worth noting that the *same* model (namely, DefaultButtonModel) is used for push buttons, radio buttons, checkboxes, and even menu items. Of course, each of these button types has different views and controllers. When using the Metal look-and-feel, the JButton uses a class called BasicButtonUI for the view and a class called ButtonUIListener as controller. In general, each Swing component has an associated view object that ends in UI. But not all Swing components have dedicated controller objects.

So, having read this short introduction to what is going on under the hood in a JButton, you may be wondering: Just what is a JButton really? It is simply a wrapper class inheriting from JComponent that holds the DefaultButtonModel object, some view data (such as the button label and icons), and a BasicButtonUI object that is responsible for the button view.

11.2 Introduction to Layout Management

Before we go on to discussing individual Swing components, such as text fields and radio buttons, we briefly cover how to arrange these components inside a frame.

Of course, Java development environments have drag-and-drop GUI builders. Nevertheless, it is important to know exactly what goes on "under the hood" because even the best of these tools will usually require hand-tweaking.

11.2.1 Layout Managers

Let's start by reviewing the program from Listing 10.4 that used buttons to change the background color of a frame.

The buttons are contained in a JPanel object and are managed by the *flow layout manager*, the default layout manager for a panel. Figure 11.4 shows what happens when you add more buttons to the panel. As you can see, a new row is started when there is no more room.

Figure 11.4 A panel with six buttons managed by a flow layout

Moreover, the buttons stay centered in the panel, even when the user resizes the frame (see Figure 11.5).

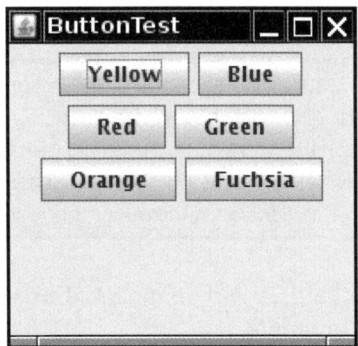

Figure 11.5 Changing the panel size rearranges the buttons automatically.

In general, *components* are placed inside *containers*, and a *layout manager* determines the positions and sizes of components in a container.

Buttons, text fields, and other user interface elements extend the class Component. Components can be placed inside containers, such as panels. Containers can themselves be put inside other containers, so the class Container extends Component. Figure 11.6 shows the inheritance hierarchy for Component.

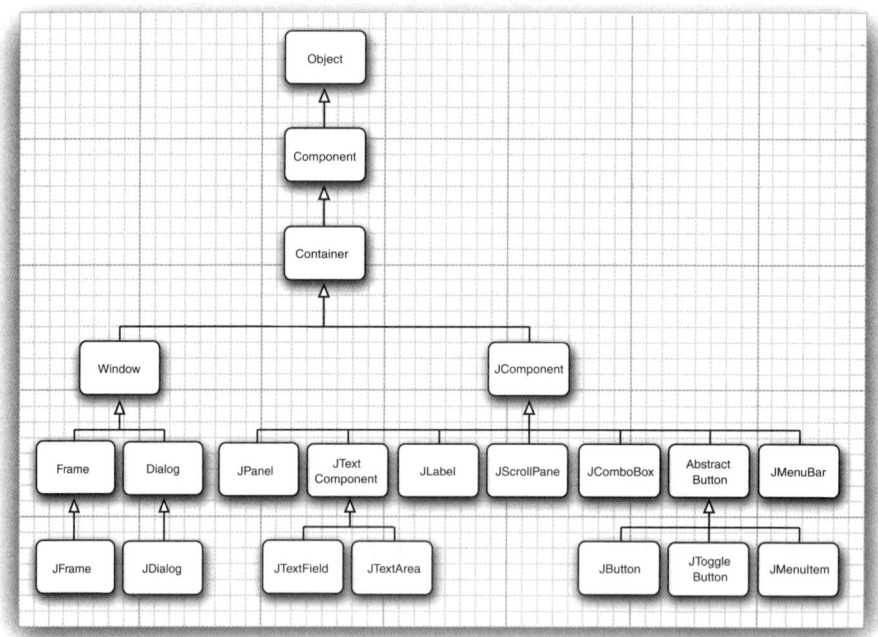

Figure 11.6 Inheritance hierarchy for the Component class

 NOTE: Unfortunately, the inheritance hierarchy is somewhat unclean in two respects. First, top-level windows, such as JFrame, are subclasses of Container and hence Component, but they cannot be placed inside other containers. Moreover, JComponent is a subclass of Container, not Component. Therefore one can add other components into a JButton. (However, those components would not be displayed.)

Each container has a default layout manager, but you can always set your own. For example, the statement

```
panel.setLayout(new GridLayout(4, 4));
```

uses the GridLayout class to lay out the components in four rows and four columns. When you add components to the container, the add method of the container passes the component and any placement directions to the layout manager.

---

java.awt.Container 1.0

- void setLayout(LayoutManager m)

  sets the layout manager for this container.

- Component add(Component c)
- Component add(Component c, Object constraints) 1.1

  adds a component to this container and returns the component reference.

---

java.awt.FlowLayout 1.0

- FlowLayout()
- FlowLayout(int align)
- FlowLayout(int align, int hgap, int vgap)

  constructs a new FlowLayout. The align parameter is one of LEFT, CENTER, or RIGHT.

11.2.2 Border Layout

The *border layout manager* is the default layout manager of the content pane of every JFrame. Unlike the flow layout manager, which completely controls the position of each component, the border layout manager lets you choose where you want to place each component. You can choose to place the component in the center, north, south, east, or west of the content pane (see Figure 11.7).

For example:

```
frame.add(component, BorderLayout.SOUTH);
```

The edge components are laid out first, and the remaining available space is occupied by the center. When the container is resized, the dimensions of the edge components are unchanged, but the center component changes its size. Add components by specifying a constant CENTER, NORTH, SOUTH, EAST, or WEST of the BorderLayout class. Not all of the positions need to be occupied. If you don't supply any value, CENTER is assumed.

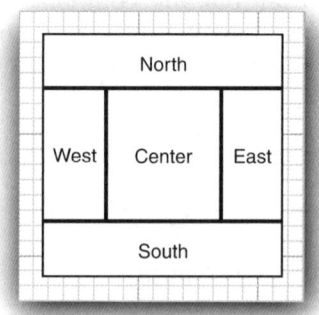

Figure 11.7 Border layout

 NOTE: The BorderLayout constants are defined as strings. For example, BorderLayout.SOUTH is defined as the string "South". This is safer than using strings. If you accidentally misspell a string, for example, frame.add(component, "south"), the compiler won't catch that error.

Unlike the flow layout, the border layout grows all components to fill the available space. (The flow layout leaves each component at its preferred size.) This is a problem when you add a button:

```
frame.add(yellowButton, BorderLayout.SOUTH); // don't
```

Figure 11.8 shows what happens when you use the preceding code fragment. The button has grown to fill the entire southern region of the frame. And, if you were to add another button to the southern region, it would just displace the first button.

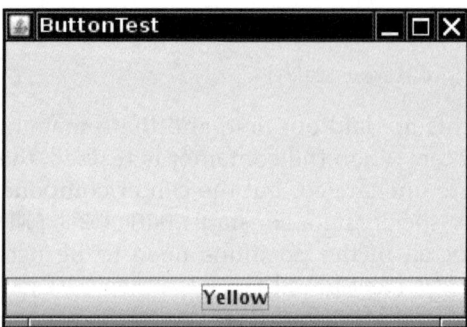

Figure 11.8 A single button managed by a border layout

To solve this problem, use additional panels. For example, look at Figure 11.9. The three buttons at the bottom of the screen are all contained in a panel. The panel is put into the southern region of the content pane.

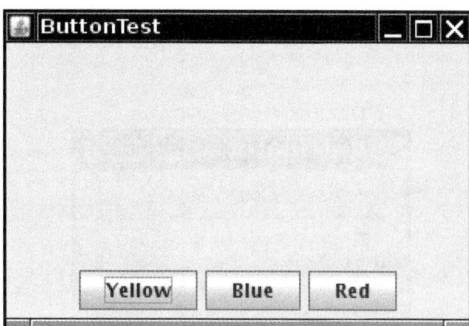

Figure 11.9 Panel placed at the southern region of the frame

To achieve this configuration, first create a new JPanel object, then add the individual buttons to the panel. The default layout manager for a panel is a FlowLayout, which is a good choice for this situation. Add the individual buttons to the panel, using the add method you have seen before. The position and size of the buttons is under the control of the FlowLayout manager. This means the buttons stay centered within the panel and do not expand to fill the entire panel area. Finally, add the panel to the content pane of the frame.

```
var panel = new JPanel();
panel.add(yellowButton);
panel.add(blueButton);
panel.add(redButton);
frame.add(panel, BorderLayout.SOUTH);
```

The border layout expands the size of the panel to fill the entire southern region.

java.awt.BorderLayout   1.0
• BorderLayout()   • BorderLayout(int hgap, int vgap)   constructs a new BorderLayout.

11.2.3 Grid Layout

The grid layout arranges all components in rows and columns like a spreadsheet. All components are given the same size. The calculator program in Figure 11.10 uses a grid layout to arrange the calculator buttons. When you resize the window, the buttons grow and shrink, but all buttons have identical sizes.

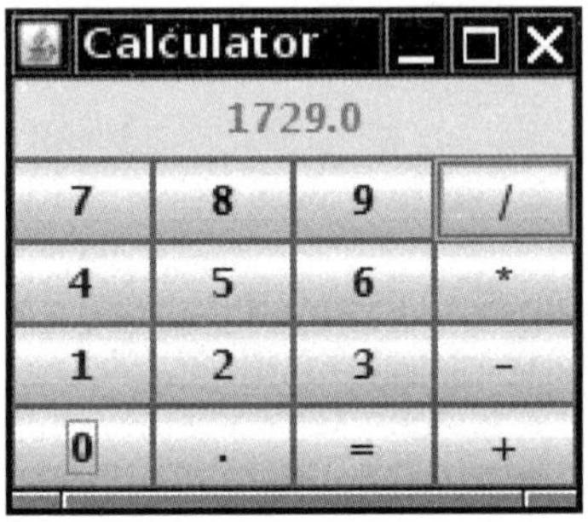

Figure 11.10 A calculator

In the constructor of the grid layout object, you specify how many rows and columns you need.

```
panel.setLayout(new GridLayout(4, 4));
```

Add the components, starting with the first entry in the first row, then the second entry in the first row, and so on.

```
panel.add(new JButton("1"));
panel.add(new JButton("2"));
```

Of course, few applications have as rigid a layout as the face of a calculator. In practice, small grids (usually with just one row or one column) can be useful to organize partial areas of a window. For example, if you want to have a row of buttons of identical sizes, you can put the buttons inside a panel that is governed by a grid layout with a single row.

---

java.awt.GridLayout 1.0

- GridLayout(int rows, int columns)
- GridLayout(int rows, int columns, int hgap, int vgap)

  constructs a new GridLayout. One of rows and columns (but not both) may be zero, denoting an arbitrary number of components per row or column.

11.3 Text Input

We are finally ready to start introducing the Swing user interface components. We begin with the components that let a user input and edit text. You can use the JTextField and JTextArea components for text input. A text field can accept only one line of text; a text area can accept multiple lines of text. A JPasswordField accepts one line of text without showing the contents.

All three of these classes inherit from a class called JTextComponent. You will not be able to construct a JTextComponent yourself because it is an abstract class. On the other hand, as is so often the case in Java, when you go searching through the API documentation, you may find that the methods you are looking for are actually in the parent class JTextComponent rather than the derived class. For example, the methods that get or set the text in a text field or text area are actually in JTextComponent.

---

javax.swing.text.JTextComponent 1.2

- String getText()
- void setText(String text)

  gets or sets the text of this text component.

- boolean isEditable()
- void setEditable(boolean b)

  gets or sets the editable property that determines whether the user can edit the content of this text component.

11.3.1 Text Fields

The usual way to add a text field to a window is to add it to a panel or other container—just as you would add a button:

```
var panel = new JPanel();
var textField = new JTextField("Default input", 20);
panel.add(textField);
```

This code adds a text field and initializes it by placing the string "Default input" inside it. The second parameter of this constructor sets the width. In this case, the width is 20 "columns." Unfortunately, a column is a rather imprecise measurement. One column is the expected width of one character in the font you are using for the text. The idea is that if you expect the inputs to be n characters or less, you are supposed to specify n as the column width. In practice, this measurement doesn't work out too well, and you should add 1 or 2 to the maximum input length to be on the safe side. Also, keep in mind that the number of columns is only a hint to the AWT that gives the *preferred* size. If the layout manager needs to grow or shrink the text field, it can adjust its size. The column width that you set in the JTextField constructor is not an upper limit on the number of characters the user can enter. The user can still type in longer strings, but the input scrolls when the text exceeds the length of the field. Users tend to find scrolling text fields irritating, so you should size the fields generously. If you need to reset the number of columns at runtime, you can do that with the setColumns method.

 TIP: After changing the size of a text box with the setColumns method, call the revalidate method of the surrounding container.

```
textField.setColumns(10);
panel.revalidate();
```

The revalidate method recomputes the size and layout of all components in a container. After you use the revalidate method, the layout manager resizes the container, and the changed size of the text field will be visible.

The revalidate method belongs to the JComponent class. It doesn't immediately resize the component but merely marks it for resizing. This approach avoids repetitive calculations if multiple components request to be resized. However, if you want to recompute all components inside a JFrame, you have to call the validate method—JFrame doesn't extend JComponent.

In general, users add text (or edit an existing text) in a text field. Quite often these text fields start out blank. To make a blank text field, just leave out the string as a parameter for the JTextField constructor:

```
var textField = new JTextField(20);
```

You can change the content of the text field at any time by using the setText method from the JTextComponent parent class mentioned in the previous section. For example:

```
textField.setText("Hello!");
```

And, as was mentioned in the previous section, you can find out what the user typed by calling the getText method. This method returns the exact text that the user has typed. To trim any extraneous leading and trailing spaces from the data in a text field, apply the trim method to the return value of getText:

```
String text = textField.getText().trim();
```

To change the font in which the user text appears, use the setFont method.

---

javax.swing.JTextField 1.2

- JTextField(int cols)

  constructs an empty JTextField with the specified number of columns.
- JTextField(String text, int cols)

  constructs a new JTextField with an initial string and the specified number of columns.
- int getColumns()
- void setColumns(int cols)

  gets or sets the number of columns that this text field should use.

---

javax.swing.JComponent 1.2

- void revalidate()

  causes the position and size of a component to be recomputed.
- void setFont(Font f)

  sets the font of this component.

---

java.awt.Component 1.0

- void validate()

  recomputes the position and size of a component. If the component is a container, the positions and sizes of its components are recomputed.
- Font getFont()

  gets the font of this component.

11.3.2 Labels and Labeling Components

Labels are components that hold text. They have no decorations (for example, no boundaries). They also do not react to user input. You can use a label to identify components. For example, unlike buttons, text fields have no label to identify them. To label a component that does not itself come with an identifier:

1. Construct a JLabel component with the correct text.
2. Place it close enough to the component you want to identify so that the user can see that the label identifies the correct component.

The constructor for a JLabel lets you specify the initial text or icon and, optionally, the alignment of the content. Use constants from the SwingConstants interface to specify alignment. That interface defines a number of useful constants such as LEFT, RIGHT, CENTER, NORTH, EAST, and so on. The JLabel class is one of several Swing classes that implement this interface. Therefore, you can specify a right-aligned label either as

```
var label = new JLabel("User name: ", SwingConstants.RIGHT);
```

or

```
var label = new JLabel("User name: ", JLabel.RIGHT);
```

The setText and setIcon methods let you set the text and icon of the label at runtime.

 TIP: You can use both plain and HTML text in buttons, labels, and menu items. We don't recommend HTML in buttons—it interferes with the look-and-feel. But HTML in labels can be very effective. Simply surround the label string with <html>. . .</html>, like this:

```
label = new JLabel("<html><b>Required</b> entry:</html>");
```

Note that the first component with an HTML label may take some time to be displayed because the rather complex HTML rendering code must be loaded.

Labels can be positioned inside a container like any other component. This means you can use the techniques you have seen before to place your labels where you need them.

---

javax.swing.JLabel 1.2

- JLabel(String text)
- JLabel(Icon icon)
- JLabel(String text, int align)
- JLabel(String text, Icon icon, int align)

  constructs a label. The align parameter is one of the SwingConstants constants LEFT (default), CENTER, or RIGHT.

- String getText()
- void setText(String text)

  gets or sets the text of this label.

- Icon getIcon()
- void setIcon(Icon icon)

  gets or sets the icon of this label.

11.3.3 Password Fields

Password fields are a special kind of text fields. To prevent nosy bystanders from seeing your password, the characters that the user enters are not actually displayed. Instead, each typed character is represented by an *echo character*, such as a bullet (·). Swing supplies a JPasswordField class that implements such a text field.

The password field is another example of the power of the model-view-controller architecture pattern. The password field uses the same model to store the data as a regular text field, but its view has been changed to display all characters as echo characters.

---

javax.swing.JPasswordField 1.2

- JPasswordField(String text, int columns)

  constructs a new password field.

- void setEchoChar(char echo)

  sets the echo character for this password field. This is advisory; a particular look-and-feel may insist on its own choice of echo character. A value of 0 resets the echo character to the default.

- char[] getPassword()

  returns the text contained in this password field. For stronger security, you should overwrite the content of the returned array after use. (The password is not returned as a String because a string would stay in the virtual machine until it is garbage-collected.)

11.3.4 Text Areas

Sometimes, you need to collect user input that is more than one line long. As mentioned earlier, you can use the JTextArea component for this. When you place a text area component in your program, a user can enter any number of lines of text, using the Enter key to separate them. Each line ends with a '\n'. Figure 11.11 shows a text area at work.

In the constructor for the JTextArea component, specify the number of rows and columns for the text area. For example,

```
textArea = new JTextArea(8, 40); // 8 lines of 40 columns each
```

where the columns parameter works as before—and you still need to add a few more columns for safety's sake. Also, as before, the user is not restricted to the number of rows and columns; the text simply scrolls when the user inputs

Figure 11.11 Text components

too much. You can also use the setColumns method to change the number of columns and the setRows method to change the number of rows. These numbers only indicate the preferred size—the layout manager can still grow or shrink the text area.

If there is more text than the text area can display, the remaining text is simply clipped. You can avoid clipping long lines by turning on line wrapping:

```
textArea.setLineWrap(true); // long lines are wrapped
```

This wrapping is a visual effect only; the text in the document is not changed—no automatic '\n' characters are inserted into the text.

11.3.5 Scroll Panes

In Swing, a text area does not have scrollbars. If you want scrollbars, you have to place the text area inside a *scroll pane*.

```
textArea = new JTextArea(8, 40);
var scrollPane = new JScrollPane(textArea);
```

The scroll pane now manages the view of the text area. Scrollbars automatically appear if there is more text than the text area can display, and they vanish again if text is deleted and the remaining text fits inside the area. The scrolling is handled internally by the scroll pane—your program does not need to process scroll events.

This is a general mechanism that works for any component, not just text areas. To add scrollbars to a component, put them inside a scroll pane.

Listing 11.1 demonstrates the various text components. This program shows a text field, a password field, and a text area with scrollbars. The text field and password field are labeled. Click on "Insert" to insert the field contents into the text area.

 NOTE: The JTextArea component displays plain text only, without special fonts or formatting. To display formatted text (such as HTML), you can use the JEditorPane class that is discussed in Volume II.

Listing 11.1 text/TextComponentFrame.java

```java
package text;

import java.awt.BorderLayout;
import java.awt.GridLayout;

import javax.swing.JButton;
import javax.swing.JFrame;
import javax.swing.JLabel;
import javax.swing.JPanel;
import javax.swing.JPasswordField;
import javax.swing.JScrollPane;
import javax.swing.JTextArea;
import javax.swing.JTextField;
import javax.swing.SwingConstants;

/**
 * A frame with sample text components.
 */
public class TextComponentFrame extends JFrame
{
   public static final int TEXTAREA_ROWS = 8;
   public static final int TEXTAREA_COLUMNS = 20;

   public TextComponentFrame()
   {
      var textField = new JTextField();
      var passwordField = new JPasswordField();

      var northPanel = new JPanel();
      northPanel.setLayout(new GridLayout(2, 2));
      northPanel.add(new JLabel("User name: ", SwingConstants.RIGHT));
```

(Continues)

Listing 11.1 *(Continued)*

```
32      northPanel.add(textField);
33      northPanel.add(new JLabel("Password: ", SwingConstants.RIGHT));
34      northPanel.add(passwordField);
35
36      add(northPanel, BorderLayout.NORTH);
37
38      var textArea = new JTextArea(TEXTAREA_ROWS, TEXTAREA_COLUMNS);
39      var scrollPane = new JScrollPane(textArea);
40
41      add(scrollPane, BorderLayout.CENTER);
42
43      // add button to append text into the text area
44
45      var southPanel = new JPanel();
46
47      var insertButton = new JButton("Insert");
48      southPanel.add(insertButton);
49      insertButton.addActionListener(event ->
50        textArea.append("User name: " + textField.getText() + " Password: "
51          + new String(passwordField.getPassword()) + "\n"));
52
53      add(southPanel, BorderLayout.SOUTH);
54      pack();
55    }
56 }
```

javax.swing.JTextArea 1.2

- JTextArea()
- JTextArea(int rows, int cols)
- JTextArea(String text, int rows, int cols)

  constructs a new text area.

- void setColumns(int cols)

  tells the text area the preferred number of columns it should use.

- void setRows(int rows)

  tells the text area the preferred number of rows it should use.

- void append(String newText)

  appends the given text to the end of the text already in the text area.

- void setLineWrap(boolean wrap)

  turns line wrapping on or off.

(Continues)

---

javax.swing.JTextArea 1.2 *(Continued)*

- void setWrapStyleWord(boolean word)

  If word is true, long lines are wrapped at word boundaries. If it is false, long lines are broken without taking word boundaries into account.

- void setTabSize(int c)

  sets tab stops every c columns. Note that the tabs aren't converted to spaces but cause alignment with the next tab stop.

---

javax.swing.JScrollPane 1.2

- JScrollPane(Component c)

  creates a scroll pane that displays the content of the specified component. Scrollbars are supplied when the component is larger than the view.

11.4 Choice Components

You now know how to collect text input from users, but there are many occasions where you would rather give users a finite set of choices than have them enter the data in a text component. Using a set of buttons or a list of items tells your users what choices they have. (It also saves you the trouble of error checking.) In this section, you will learn how to program checkboxes, radio buttons, lists of choices, and sliders.

11.4.1 Checkboxes

If you want to collect just a "yes" or "no" input, use a checkbox component. Checkboxes automatically come with labels that identify them. The user can check the box by clicking inside it and turn off the checkmark by clicking inside the box again. Pressing the space bar when the focus is in the checkbox also toggles the checkmark.

Figure 11.12 shows a simple program with two checkboxes, one for turning the italic attribute of a font on or off, and the other for boldface. Note that the second checkbox has focus, as indicated by the rectangle around the label. Each time the user clicks one of the checkboxes, the screen is refreshed, using the new font attributes.

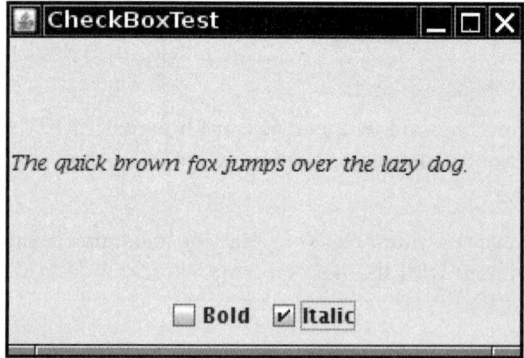

Figure 11.12 Checkboxes

Checkboxes need a label next to them to identify their purpose. Give the label text in the constructor:

```
bold = new JCheckBox("Bold");
```

Use the setSelected method to turn a checkbox on or off. For example:

```
bold.setSelected(true);
```

The isSelected method then retrieves the current state of each checkbox. It is false if unchecked, true if checked.

When the user clicks on a checkbox, this triggers an action event. As always, you attach an action listener to the checkbox. In our program, the two checkboxes share the same action listener.

```
ActionListener listener = . . .;
bold.addActionListener(listener);
italic.addActionListener(listener);
```

The listener queries the state of the bold and italic checkboxes and sets the font of the panel to plain, bold, italic, or both bold and italic.

```
ActionListener listener = event -> {
    int mode = 0;
    if (bold.isSelected()) mode += Font.BOLD;
    if (italic.isSelected()) mode += Font.ITALIC;
    label.setFont(new Font(Font.SERIF, mode, FONTSIZE));
};
```

Listing 11.2 is the program listing for the checkbox example.

Listing 11.2 checkBox/CheckBoxFrame.java

```java
1  package checkBox;
2
3  import java.awt.*;
4  import java.awt.event.*;
5  import javax.swing.*;
6
7  /**
8   * A frame with a sample text label and check boxes for selecting font
9   * attributes.
10  */
11 public class CheckBoxFrame extends JFrame
12 {
13     private JLabel label;
14     private JCheckBox bold;
15     private JCheckBox italic;
16     private static final int FONTSIZE = 24;
17
18     public CheckBoxFrame()
19     {
20         // add the sample text label
21
22         label = new JLabel("The quick brown fox jumps over the lazy dog.");
23         label.setFont(new Font("Serif", Font.BOLD, FONTSIZE));
24         add(label, BorderLayout.CENTER);
25
26         // this listener sets the font attribute of
27         // the label to the check box state
28
29         ActionListener listener = event -> {
30             int mode = 0;
31             if (bold.isSelected()) mode += Font.BOLD;
32             if (italic.isSelected()) mode += Font.ITALIC;
33             label.setFont(new Font("Serif", mode, FONTSIZE));
34         };
35
36         // add the check boxes
37
38         var buttonPanel = new JPanel();
39
40         bold = new JCheckBox("Bold");
41         bold.addActionListener(listener);
42         bold.setSelected(true);
43         buttonPanel.add(bold);
44
45         italic = new JCheckBox("Italic");
46         italic.addActionListener(listener);
47         buttonPanel.add(italic);
```

(Continues)

Listing 11.2 *(Continued)*

```
48
49      add(buttonPanel, BorderLayout.SOUTH);
50      pack();
51   }
52 }
```

---

javax.swing.JCheckBox 1.2

- JCheckBox(String label)
- JCheckBox(String label, Icon icon)

  constructs a checkbox that is initially unselected.

- JCheckBox(String label, boolean state)

  constructs a checkbox with the given label and initial state.

- boolean isSelected()
- void setSelected(boolean state)

  gets or sets the selection state of the checkbox.

11.4.2 Radio Buttons

In the previous example, the user could check either, both, or neither of the two checkboxes. In many cases, we want the user to check only one of several boxes. When another box is checked, the previous box is automatically unchecked. Such a group of boxes is often called a *radio button group* because the buttons work like the station selector buttons on a radio. When you push in one button, the previously depressed button pops out. Figure 11.13 shows a typical example. We allow the user to select a font size from among the choices—Small, Medium, Large, or Extra large—but, of course, we will allow selecting only one size at a time.

Implementing radio button groups is easy in Swing. You construct one object of type ButtonGroup for every group of buttons. Then, you add objects of type JRadioButton to the button group. The button group object is responsible for turning off the previously set button when a new button is clicked.

```
var group = new ButtonGroup();

var smallButton = new JRadioButton("Small", false);
group.add(smallButton);

var mediumButton = new JRadioButton("Medium", true);
group.add(mediumButton);
. . .
```

Figure 11.13 A radio button group

The second argument of the constructor is true for the button that should be checked initially and false for all others. Note that the button group controls only the *behavior* of the buttons; if you want to group the buttons for layout purposes, you also need to add them to a container such as a JPanel.

If you look again at Figures 11.12 and 11.13, you will note that the appearance of the radio buttons is different from that of checkboxes. Checkboxes are square and contain a checkmark when selected. Radio buttons are round and contain a dot when selected.

The event notification mechanism for radio buttons is the same as for any other buttons. When the user checks a radio button, the button generates an action event. In our example program, we define an action listener that sets the font size to a particular value:

```
ActionListener listener = event ->
    label.setFont(new Font("Serif", Font.PLAIN, size));
```

Compare this listener setup to that of the checkbox example. Each radio button gets a different listener object. Each listener object knows exactly what it needs to do—set the font size to a particular value. With checkboxes, we used a different approach: Both checkboxes have the same action listener that calls a method looking at the current state of both checkboxes.

Could we follow the same approach here? We could have a single listener that computes the size as follows:

```
if (smallButton.isSelected()) size = 8;
else if (mediumButton.isSelected()) size = 12;
. . .
```

However, we prefer to use separate action listener objects because they tie the size values more closely to the buttons.

NOTE: If you have a group of radio buttons, you know that only one of them is selected. It would be nice to be able to quickly find out which, without having to query all the buttons in the group. The ButtonGroup object controls all buttons, so it would be convenient if this object could give us a reference to the selected button. Indeed, the ButtonGroup class has a getSelection method, but that method doesn't return the radio button that is selected. Instead, it returns a ButtonModel reference to the model attached to the button. Unfortunately, none of the ButtonModel methods are very helpful. The ButtonModel interface inherits a method getSelectedObjects from the ItemSelectable interface that, rather uselessly, returns null. The getActionCommand method looks promising because the "action command" of a radio button is its text label. But the action command of its model is null. Only if you explicitly set the action commands of all radio buttons with the setActionCommand method do the action command values of the models also get set. Then you can retrieve the action command of the currently selected button with buttonGroup.getSelection().getActionCommand().

Listing 11.3 is the complete program for font size selection that puts a set of radio buttons to work.

Listing 11.3 radioButton/RadioButtonFrame.java

```java
1  package radioButton;
2
3  import java.awt.*;
4  import java.awt.event.*;
5  import javax.swing.*;
6
7  /**
8   * A frame with a sample text label and radio buttons for selecting font sizes.
9   */
10 public class RadioButtonFrame extends JFrame
11 {
12    private JPanel buttonPanel;
13    private ButtonGroup group;
14    private JLabel label;
15    private static final int DEFAULT_SIZE = 36;
16
17    public RadioButtonFrame()
18    {
19       // add the sample text label
20
21       label = new JLabel("The quick brown fox jumps over the lazy dog.");
22       label.setFont(new Font("Serif", Font.PLAIN, DEFAULT_SIZE));
```

```
23        add(label, BorderLayout.CENTER);
24
25        // add the radio buttons
26
27        buttonPanel = new JPanel();
28        group = new ButtonGroup();
29
30        addRadioButton("Small", 8);
31        addRadioButton("Medium", 12);
32        addRadioButton("Large", 18);
33        addRadioButton("Extra large", 36);
34
35        add(buttonPanel, BorderLayout.SOUTH);
36        pack();
37     }
38
39     /**
40      * Adds a radio button that sets the font size of the sample text.
41      * @param name the string to appear on the button
42      * @param size the font size that this button sets
43      */
44     public void addRadioButton(String name, int size)
45     {
46        boolean selected = size == DEFAULT_SIZE;
47        var button = new JRadioButton(name, selected);
48        group.add(button);
49        buttonPanel.add(button);
50
51        // this listener sets the label font size
52
53        ActionListener listener = event -> label.setFont(new Font("Serif", Font.PLAIN, size));
54
55        button.addActionListener(listener);
56     }
57 }
```

---

javax.swing.JRadioButton 1.2

- JRadioButton(String label, Icon icon)

  constructs a radio button that is initially unselected.

- JRadioButton(String label, boolean state)

  constructs a radio button with the given label and initial state.

---

javax.swing.ButtonGroup 1.2

- void add(AbstractButton b)

  adds the button to the group.
- ButtonModel getSelection()

  returns the button model of the selected button.

---

javax.swing.ButtonModel 1.2

- String getActionCommand()

  returns the action command for this button model.

---

javax.swing.AbstractButton 1.2

- void setActionCommand(String s)

  sets the action command for this button and its model.

11.4.3 Borders

If you have multiple groups of radio buttons in a window, you will want to visually indicate which buttons are grouped. Swing provides a set of useful *borders* for this purpose. You can apply a border to any component that extends JComponent. The most common usage is to place a border around a panel and fill that panel with other user interface elements, such as radio buttons.

You can choose from quite a few borders, but you need to follow the same steps for all of them.

1. Call a static method of the BorderFactory to create a border. You can choose among the following styles (see Figure 11.14):

    - Lowered bevel
    - Raised bevel
    - Etched
    - Line
    - Matte
    - Empty (just to create some blank space around the component)

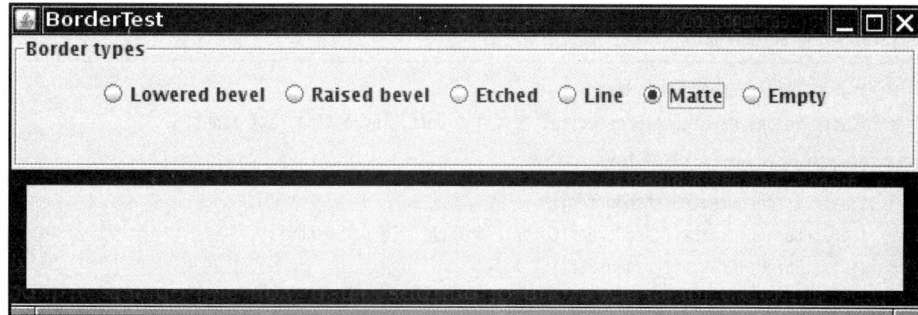

Figure 11.14 Testing border types

2. If you like, add a title to your border by passing your border to BorderFactory
   .createTitledBorder.

3. If you really want to go all out, combine several borders with a call to
   BorderFactory.createCompoundBorder.

4. Add the resulting border to your component by calling the setBorder method
   of the JComponent class.

For example, here is how you add an etched border with a title to a panel:

```
Border etched = BorderFactory.createEtchedBorder();
Border titled = BorderFactory.createTitledBorder(etched, "A Title");
panel.setBorder(titled);
```

Different borders have different options for setting border widths and colors;
see the API notes for details. True border enthusiasts will appreciate that
there is also a SoftBevelBorder class for beveled borders with softened corners
and that a LineBorder can have rounded corners as well. You can construct these
borders only by using one of the class constructors—there is no BorderFactory
method for them.

---

javax.swing.BorderFactory 1.2

- static Border createLineBorder(Color color)
- static Border createLineBorder(Color color, int thickness)

  creates a simple line border.

- static MatteBorder createMatteBorder(int top, int left, int bottom, int right, Color color)
- static MatteBorder createMatteBorder(int top, int left, int bottom, int right, Icon tileIcon)

  creates a thick border that is filled with a color or a repeating icon.

(Continues)

---

javax.swing.BorderFactory 1.2 *(Continued)*

- static Border createEmptyBorder()
- static Border createEmptyBorder(int top, int left, int bottom, int right)

creates an empty border.

- static Border createEtchedBorder()
- static Border createEtchedBorder(Color highlight, Color shadow)
- static Border createEtchedBorder(int type)
- static Border createEtchedBorder(int type, Color highlight, Color shadow)

creates a line border with a 3D effect. The type parameter is one of EtchedBorder
.RAISED, EtchedBorder.LOWERED.

- static Border createBevelBorder(int type)
- static Border createBevelBorder(int type, Color highlight, Color shadow)
- static Border createLoweredBevelBorder()
- static Border createRaisedBevelBorder()

creates a border that gives the effect of a lowered or raised surface. The type
parameter is one of BevelBorder.RAISED, BevelBorder.LOWERED.

- static TitledBorder createTitledBorder(String title)
- static TitledBorder createTitledBorder(Border border)
- static TitledBorder createTitledBorder(Border border, String title)
- static TitledBorder createTitledBorder(Border border, String title, int justification, int position)
- static TitledBorder createTitledBorder(Border border, String title, int justification, int position, Font font)
- static TitledBorder createTitledBorder(Border border, String title, int justification, int position, Font font, Color color)

creates a titled border with the specified properties. The justification parameter
is one of the TitledBorder constants LEFT, CENTER, RIGHT, LEADING, TRAILING, or
DEFAULT_JUSTIFICATION (left), and position is one of ABOVE_TOP, TOP, BELOW_TOP, ABOVE_BOTTOM,
BOTTOM, BELOW_BOTTOM, or DEFAULT_POSITION (top).

- static CompoundBorder createCompoundBorder(Border outsideBorder, Border insideBorder)

combines two borders to a new border.

---

javax.swing.border.SoftBevelBorder 1.2

- SoftBevelBorder(int type)
- SoftBevelBorder(int type, Color highlight, Color shadow)

creates a bevel border with softened corners. The type parameter is one of
SoftBevelBorder.RAISED, SoftBevelBorder.LOWERED.

---

javax.swing.border.LineBorder 1.2

* public LineBorder(Color color, int thickness, boolean roundedCorners)

  creates a line border with the given color and thickness. If roundedCorners is true, the border has rounded corners.

---

javax.swing.JComponent 1.2

* void setBorder(Border border)

  sets the border of this component.

11.4.4 Combo Boxes

If you have more than a handful of alternatives, radio buttons are not a good choice because they take up too much screen space. Instead, you can use a combo box. When the user clicks on this component, a list of choices drops down, and the user can then select one of them (see Figure 11.15).

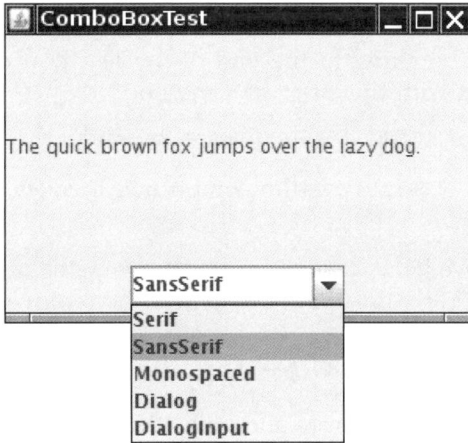

Figure 11.15 A combo box

If the drop-down list box is set to be *editable*, you can edit the current selection as if it were a text field. For that reason, this component is called a *combo box*—it combines the flexibility of a text field with a set of predefined choices. The JComboBox class provides a combo box component.

As of Java 7, the JComboBox class is a generic class. For example, a JComboBox<String> holds objects of type String, and a JComboBox<Integer> holds integers.

Call the setEditable method to make the combo box editable. Note that editing affects only the selected item. It does not change the list of choices in any way.

You can obtain the current selection, which may have been edited if the combo box is editable, by calling the getSelectedItem method. However, for an editable combo box, that item may have any type, depending on the editor that takes the user edits and turns the result into an object. (See Volume II, Chapter 6 for a discussion of editors.) If your combo box isn't editable, you are better off calling

```
combo.getItemAt(combo.getSelectedIndex())
```

which gives you the selected item with the correct type.

In the example program, the user can choose a font style from a list of styles (Serif, SansSerif, Monospaced, etc.). The user can also type in another font.

Add the choice items with the addItem method. In our program, addItem is called only in the constructor, but you can call it any time.

```
var faceCombo = new JComboBox<String>();
faceCombo.addItem("Serif");
faceCombo.addItem("SansSerif");
. . .
```

This method adds the string to the end of the list. You can add new items anywhere in the list with the insertItemAt method:

```
faceCombo.insertItemAt("Monospaced", 0); // add at the beginning
```

You can add items of any type—the combo box invokes each item's toString method to display it.

If you need to remove items at runtime, use the removeItem or removeItemAt method, depending on whether you supply the item to be removed or its position.

```
faceCombo.removeItem("Monospaced");
faceCombo.removeItemAt(0); // remove first item
```

The removeAllItems method removes all items at once.

 TIP: If you need to add a large number of items to a combo box, the addItem method will perform poorly. Instead, construct a DefaultComboBoxModel, populate it by calling addElement, and then call the setModel method of the JComboBox class.

When the user selects an item from a combo box, the combo box generates an action event. To find out which item was selected, call getSource on the event parameter to get a reference to the combo box that sent the event.

Then call the getSelectedItem method to retrieve the currently selected item. You will need to cast the returned value to the appropriate type, usually String.

```
ActionListener listener = event ->
   label.setFont(new Font(
      faceCombo.getItemAt(faceCombo.getSelectedIndex()),
      Font.PLAIN,
      DEFAULT_SIZE));
```

Listing 11.4 shows the complete program.

Listing 11.4 comboBox/ComboBoxFrame.java

```
1  package comboBox;
2
3  import java.awt.BorderLayout;
4  import java.awt.Font;
5
6  import javax.swing.JComboBox;
7  import javax.swing.JFrame;
8  import javax.swing.JLabel;
9  import javax.swing.JPanel;
10
11 /**
12  * A frame with a sample text label and a combo box for selecting font faces.
13  */
14 public class ComboBoxFrame extends JFrame
15 {
16    private JComboBox<String> faceCombo;
17    private JLabel label;
18    private static final int DEFAULT_SIZE = 24;
19
20    public ComboBoxFrame()
21    {
22       // add the sample text label
23
24       label = new JLabel("The quick brown fox jumps over the lazy dog.");
25       label.setFont(new Font("Serif", Font.PLAIN, DEFAULT_SIZE));
26       add(label, BorderLayout.CENTER);
27
28       // make a combo box and add face names
29
30       faceCombo = new JComboBox<>();
31       faceCombo.addItem("Serif");
32       faceCombo.addItem("SansSerif");
33       faceCombo.addItem("Monospaced");
34       faceCombo.addItem("Dialog");
35       faceCombo.addItem("DialogInput");
```

(Continues)

Listing 11.4 *(Continued)*

```
36
37      // the combo box listener changes the label font to the selected face name
38
39      faceCombo.addActionListener(event ->
40        label.setFont(
41          new Font(faceCombo.getItemAt(faceCombo.getSelectedIndex()),
42            Font.PLAIN, DEFAULT_SIZE)));
43
44      // add combo box to a panel at the frame's southern border
45
46      var comboPanel = new JPanel();
47      comboPanel.add(faceCombo);
48      add(comboPanel, BorderLayout.SOUTH);
49      pack();
50    }
51  }
```

---

javax.swing.JComboBox 1.2

- boolean isEditable()
- void setEditable(boolean b)

  gets or sets the editable property of this combo box.

- void addItem(Object item)

  adds an item to the item list.

- void insertItemAt(Object item, int index)

  inserts an item into the item list at a given index.

- void removeItem(Object item)

  removes an item from the item list.

- void removeItemAt(int index)

  removes the item at an index.

- void removeAllItems()

  removes all items from the item list.

- Object getSelectedItem()

  returns the currently selected item.

11.4.5 Sliders

Combo boxes let users choose from a discrete set of values. Sliders offer a choice from a continuum of values—for example, any number between 1 and 100.

The most common way of constructing a slider is as follows:

```
var slider = new JSlider(min, max, initialValue);
```

If you omit the minimum, maximum, and initial values, they are initialized with 0, 100, and 50, respectively.

Or if you want the slider to be vertical, use the following constructor call:

```
var slider = new JSlider(SwingConstants.VERTICAL, min, max, initialValue);
```

These constructors create a plain slider, such as the top slider in Figure 11.16. You will see presently how to add decorations to a slider.

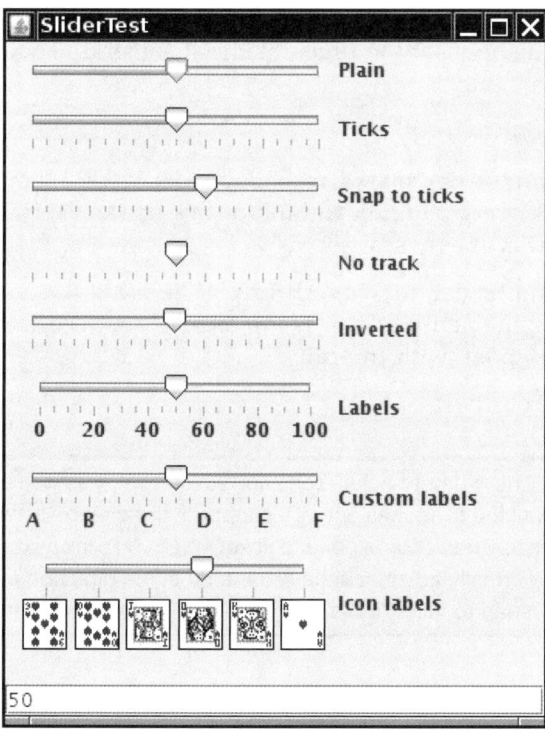

Figure 11.16 Sliders

As the user slides the slider bar, the *value* of the slider moves between the minimum and the maximum values. When the value changes, a ChangeEvent is sent to all change listeners. To be notified of the change, call the addChangeListener method and install an object that implements the functional ChangeListener interface. In the callback, retrieve the slider value:

```
ChangeListener listener = event -> {
    JSlider slider = (JSlider) event.getSource();
    int value = slider.getValue();
        . . .
};
```

You can embellish the slider by showing *ticks*. For example, in the sample program, the second slider uses the following settings:

```
slider.setMajorTickSpacing(20);
slider.setMinorTickSpacing(5);
```

The slider is decorated with large tick marks every 20 units and small tick marks every 5 units. The units refer to slider values, not pixels.

These instructions only set the units for the tick marks. To actually have the tick marks appear, call

```
slider.setPaintTicks(true);
```

The major and minor tick marks are independent. For example, you can set major tick marks every 20 units and minor tick marks every 7 units, but that will give you a very messy scale.

You can force the slider to *snap to ticks*. Whenever the user has finished dragging a slider in snap mode, it is immediately moved to the closest tick. You activate this mode with the call

```
slider.setSnapToTicks(true);
```

 CAUTION: The "snap to ticks" behavior doesn't work as well as you might imagine. Until the slider has actually snapped, the change listener still reports slider values that don't correspond to ticks. And if you click next to the slider—an action that normally advances the slider a bit in the direction of the click—a slider with "snap to ticks" does not move to the next tick.

You can display *tick mark labels* for the major tick marks by calling

```
slider.setPaintLabels(true);
```

For example, with a slider ranging from 0 to 100 and major tick spacing of 20, the ticks are labeled 0, 20, 40, 60, 80, and 100.

You can also supply other tick mark labels, such as strings or icons (see Figure 11.16). The process is a bit convoluted. You need to fill a hash table with keys of type `Integer` and values of type `Component`. You then call the `setLabelTable` method. The components are placed under the tick marks. Usually, `JLabel` objects are used. Here is how you can label ticks as A, B, C, D, E, and F:

```
var labelTable = new Hashtable<Integer, Component>();
labelTable.put(0, new JLabel("A"));
labelTable.put(20, new JLabel("B"));
  . . .
labelTable.put(100, new JLabel("F"));
slider.setLabelTable(labelTable);
```

Listing 11.5 also shows a slider with icons as tick labels.

TIP: If your tick marks or labels don't show, double-check that you called `setPaintTicks(true)` and `setPaintLabels(true)`.

The fourth slider in Figure 11.16 has no track. To suppress the "track" in which the slider moves, call

```
slider.setPaintTrack(false);
```

The fifth slider has its direction reversed by a call to

```
slider.setInverted(true);
```

The example program in Listing 11.5 shows all these visual effects with a collection of sliders. Each slider has a change event listener installed that places the current slider value into the text field at the bottom of the frame.

Listing 11.5 `slider/SliderFrame.java`

```
1 package slider;
2
3 import java.awt.*;
4 import java.util.*;
5 import javax.swing.*;
6 import javax.swing.event.*;
7
```

(Continues)

Listing 11.5 *(Continued)*

```java
 8  /**
 9   * A frame with many sliders and a text field to show slider values.
10   */
11  public class SliderFrame extends JFrame
12  {
13     private JPanel sliderPanel;
14     private JTextField textField;
15     private ChangeListener listener;
16
17     public SliderFrame()
18     {
19        sliderPanel = new JPanel();
20        sliderPanel.setLayout(new GridBagLayout());
21
22        // common listener for all sliders
23        listener = event -> {
24           // update text field when the slider value changes
25           JSlider source = (JSlider) event.getSource();
26           textField.setText("" + source.getValue());
27        };
28
29        // add a plain slider
30
31        var slider = new JSlider();
32        addSlider(slider, "Plain");
33
34        // add a slider with major and minor ticks
35
36        slider = new JSlider();
37        slider.setPaintTicks(true);
38        slider.setMajorTickSpacing(20);
39        slider.setMinorTickSpacing(5);
40        addSlider(slider, "Ticks");
41
42        // add a slider that snaps to ticks
43
44        slider = new JSlider();
45        slider.setPaintTicks(true);
46        slider.setSnapToTicks(true);
47        slider.setMajorTickSpacing(20);
48        slider.setMinorTickSpacing(5);
49        addSlider(slider, "Snap to ticks");
50
51        // add a slider with no track
52
```

```
53      slider = new JSlider();
54      slider.setPaintTicks(true);
55      slider.setMajorTickSpacing(20);
56      slider.setMinorTickSpacing(5);
57      slider.setPaintTrack(false);
58      addSlider(slider, "No track");
59
60      // add an inverted slider
61
62      slider = new JSlider();
63      slider.setPaintTicks(true);
64      slider.setMajorTickSpacing(20);
65      slider.setMinorTickSpacing(5);
66      slider.setInverted(true);
67      addSlider(slider, "Inverted");
68
69      // add a slider with numeric labels
70
71      slider = new JSlider();
72      slider.setPaintTicks(true);
73      slider.setPaintLabels(true);
74      slider.setMajorTickSpacing(20);
75      slider.setMinorTickSpacing(5);
76      addSlider(slider, "Labels");
77
78      // add a slider with alphabetic labels
79
80      slider = new JSlider();
81      slider.setPaintLabels(true);
82      slider.setPaintTicks(true);
83      slider.setMajorTickSpacing(20);
84      slider.setMinorTickSpacing(5);
85
86      var labelTable = new Hashtable<Integer, Component>();
87      labelTable.put(0, new JLabel("A"));
88      labelTable.put(20, new JLabel("B"));
89      labelTable.put(40, new JLabel("C"));
90      labelTable.put(60, new JLabel("D"));
91      labelTable.put(80, new JLabel("E"));
92      labelTable.put(100, new JLabel("F"));
93
94      slider.setLabelTable(labelTable);
95      addSlider(slider, "Custom labels");
96
97      // add a slider with icon labels
98
99      slider = new JSlider();
100     slider.setPaintTicks(true);
```

(Continues)

Listing 11.5 *(Continued)*

```
101     slider.setPaintLabels(true);
102     slider.setSnapToTicks(true);
103     slider.setMajorTickSpacing(20);
104     slider.setMinorTickSpacing(20);
105
106     labelTable = new Hashtable<Integer, Component>();
107
108     // add card images
109
110     labelTable.put(0, new JLabel(new ImageIcon("nine.gif")));
111     labelTable.put(20, new JLabel(new ImageIcon("ten.gif")));
112     labelTable.put(40, new JLabel(new ImageIcon("jack.gif")));
113     labelTable.put(60, new JLabel(new ImageIcon("queen.gif")));
114     labelTable.put(80, new JLabel(new ImageIcon("king.gif")));
115     labelTable.put(100, new JLabel(new ImageIcon("ace.gif")));
116
117     slider.setLabelTable(labelTable);
118     addSlider(slider, "Icon labels");
119
120     // add the text field that displays the slider value
121
122     textField = new JTextField();
123     add(sliderPanel, BorderLayout.CENTER);
124     add(textField, BorderLayout.SOUTH);
125     pack();
126   }
127
128   /**
129    * Adds a slider to the slider panel and hooks up the listener
130    * @param slider the slider
131    * @param description the slider description
132    */
133   public void addSlider(JSlider slider, String description)
134   {
135     slider.addChangeListener(listener);
136     var panel = new JPanel();
137     panel.add(slider);
138     panel.add(new JLabel(description));
139     panel.setAlignmentX(Component.LEFT_ALIGNMENT);
140     var gbc = new GridBagConstraints();
141     gbc.gridy = sliderPanel.getComponentCount();
142     gbc.anchor = GridBagConstraints.WEST;
143     sliderPanel.add(panel, gbc);
144   }
145 }
```

---

javax.swing.JSlider 1.2

- JSlider()
- JSlider(int direction)
- JSlider(int min, int max)
- JSlider(int min, int max, int initialValue)
- JSlider(int direction, int min, int max, int initialValue)

    constructs a horizontal slider with the given direction and minimum, maximum, and initial values. The direction parameter is one of SwingConstants.HORIZONTAL or SwingConstants.VERTICAL. The default is horizontal. Defaults for the minimum, initial, and maximum are 0, 50, and 100.

- void setPaintTicks(boolean b)

    displays ticks if b is true.

- void setMajorTickSpacing(int units)
- void setMinorTickSpacing(int units)

    sets major or minor ticks at multiples of the given slider units.

- void setPaintLabels(boolean b)

    displays tick labels if b is true.

- void setLabelTable(Dictionary table)

    sets the components to use for the tick labels. Each key/value pair in the table has the form new Integer(*value*)/*component*.

- void setSnapToTicks(boolean b)

    if b is true, then the slider snaps to the closest tick after each adjustment.

- void setPaintTrack(boolean b)

    if b is true, a track is displayed in which the slider runs.

11.5 Menus

We started this chapter by introducing the most common components that you might want to place into a window, such as various kinds of buttons, text fields, and combo boxes. Swing also supports another type of user interface element—pull-down menus that are familiar from GUI applications.

A *menu bar* at the top of a window contains the names of the pull-down menus. Clicking on a name opens the menu containing *menu items* and *submenus*. When the user clicks on a menu item, all menus are closed and a message is sent to the program. Figure 11.17 shows a typical menu with a submenu.

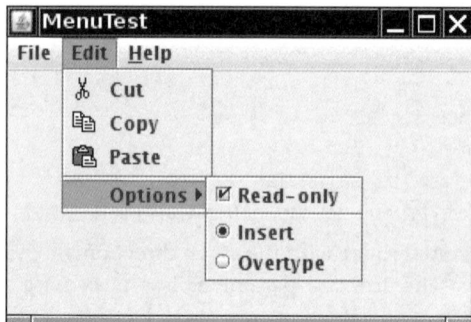

Figure 11.17 A menu with a submenu

11.5.1 Menu Building

Building menus is straightforward. First, create a menu bar:

```
var menuBar = new JMenuBar();
```

A menu bar is just a component that you can add anywhere you like. Normally, you want it to appear at the top of a frame. You can add it there with the setJMenuBar method:

```
frame.setJMenuBar(menuBar);
```

For each menu, you create a menu object:

```
var editMenu = new JMenu("Edit");
```

Add the top-level menus to the menu bar:

```
menuBar.add(editMenu);
```

Add menu items, separators, and submenus to the menu object:

```
var pasteItem = new JMenuItem("Paste");
editMenu.add(pasteItem);
editMenu.addSeparator();
JMenu optionsMenu = . . .; // a submenu
editMenu.add(optionsMenu);
```

You can see separators in Figure 11.17 below the Paste and Read-only menu items.

When the user selects a menu item, an action event is triggered. You need to install an action listener for each menu item:

```
ActionListener listener = . . .;
pasteItem.addActionListener(listener);
```

The method JMenu.add(String s) conveniently adds a menu item to the end of a menu. For example:

```
editMenu.add("Paste");
```

The add method returns the created menu item, so you can capture it and add the listener, as follows:

```
JMenuItem pasteItem = editMenu.add("Paste");
pasteItem.addActionListener(listener);
```

It often happens that menu items trigger commands that can also be activated through other user interface elements such as toolbar buttons. In Section 10.4.5, "Actions," on p. 608, you saw how to specify commands through Action objects. You define a class that implements the Action interface, usually by extending the AbstractAction convenience class, specify the menu item label in the constructor of the AbstractAction object, and override the actionPerformed method to hold the menu action handler. For example:

```
var exitAction = new AbstractAction("Exit") // menu item text goes here
   {
      public void actionPerformed(ActionEvent event)
      {
         // action code goes here
         System.exit(0);
      }
   };
```

You can then add the action to the menu:

```
JMenuItem exitItem = fileMenu.add(exitAction);
```

This command adds a menu item to the menu, using the action name. The action object becomes its listener. This is just a convenient shortcut for

```
var exitItem = new JMenuItem(exitAction);
fileMenu.add(exitItem);
```

---

javax.swing.JMenu 1.2

- JMenu(String label)

  constructs a menu with the given label.

- JMenuItem add(JMenuItem item)

  adds a menu item (or a menu).

(Continues)

---

javax.swing.JMenu 1.2 *(Continued)*

- JMenuItem add(String label)

  adds a menu item with the given label to this menu and returns the item.

- JMenuItem add(Action a)

  adds a menu item with the given action to this menu and returns the item.

- void addSeparator()

  adds a separator line to the menu.

- JMenuItem insert(JMenuItem menu, int index)

  adds a new menu item (or submenu) to the menu at a specific index.

- JMenuItem insert(Action a, int index)

  adds a new menu item with the given action at a specific index.

- void insertSeparator(int index)

  adds a separator to the menu.

- void remove(int index)
- void remove(JMenuItem item)

  removes a specific item from the menu.

---

javax.swing.JMenuItem 1.2

- JMenuItem(String label)

  constructs a menu item with a given label.

- JMenuItem(Action a) 1.3

  constructs a menu item for the given action.

---

javax.swing.AbstractButton 1.2

- void setAction(Action a) 1.3

  sets the action for this button or menu item.

---

javax.swing.JFrame 1.2

- void setJMenuBar(JMenuBar menubar)

  sets the menu bar for this frame.

11.5.2 Icons in Menu Items

Menu items are very similar to buttons. In fact, the `JMenuItem` class extends the `AbstractButton` class. Just like buttons, menus can have just a text label, just an icon, or both. You can specify the icon with the `JMenuItem(String, Icon)` or `JMenuItem(Icon)` constructor, or you can set it with the `setIcon` method that the `JMenuItem` class inherits from the `AbstractButton` class. Here is an example:

```
var cutItem = new JMenuItem("Cut", new ImageIcon("cut.gif"));
```

In Figure 11.17, you can see icons next to several menu items. By default, the menu item text is placed to the right of the icon. If you prefer the text to be placed on the left, call the `setHorizontalTextPosition` method that the `JMenuItem` class inherits from the `AbstractButton` class. For example, the call

```
cutItem.setHorizontalTextPosition(SwingConstants.LEFT);
```

moves the menu item text to the left of the icon.

You can also add an icon to an action:

```
cutAction.putValue(Action.SMALL_ICON, new ImageIcon("cut.gif"));
```

Whenever you construct a menu item out of an action, the `Action.NAME` value becomes the text of the menu item and the `Action.SMALL_ICON` value becomes the icon.

Alternatively, you can set the icon in the `AbstractAction` constructor:

```
cutAction = new
   AbstractAction("Cut", new ImageIcon("cut.gif"))
   {
      public void actionPerformed(ActionEvent event)
      {
         . . .
      }
   };
```

`javax.swing.JMenuItem` 1.2

• `JMenuItem(String label, Icon icon)`

  constructs a menu item with the given label and icon.

`javax.swing.AbstractButton` 1.2

• `void setHorizontalTextPosition(int pos)`

  sets the horizontal position of the text relative to the icon. The `pos` parameter is `SwingConstants.RIGHT` (text is to the right of icon) or `SwingConstants.LEFT`.

`javax.swing.AbstractAction` 1.2

- `AbstractAction(String name, Icon smallIcon)`

  constructs an abstract action with the given name and icon.

11.5.3 Checkbox and Radio Button Menu Items

Checkbox and *radio button* menu items display a checkbox or radio button next to the name (see Figure 11.17). When the user selects the menu item, the item automatically toggles between checked and unchecked.

Apart from the button decoration, treat these menu items just as you would any others. For example, here is how you create a checkbox menu item:

```
var readonlyItem = new JCheckBoxMenuItem("Read-only");
optionsMenu.add(readonlyItem);
```

The radio button menu items work just like regular radio buttons. You must add them to a button group. When one of the buttons in a group is selected, all others are automatically deselected.

```
var group = new ButtonGroup();
var insertItem = new JRadioButtonMenuItem("Insert");
insertItem.setSelected(true);
var overtypeItem = new JRadioButtonMenuItem("Overtype");
group.add(insertItem);
group.add(overtypeItem);
optionsMenu.add(insertItem);
optionsMenu.add(overtypeItem);
```

With these menu items, you don't necessarily want to be notified when the user selects the item. Instead, you can simply use the `isSelected` method to test the current state of the menu item. (Of course, that means you should keep a reference to the menu item stored in an instance field.) Use the `setSelected` method to set the state.

`javax.swing.JCheckBoxMenuItem` 1.2

- `JCheckBoxMenuItem(String label)`

  constructs the checkbox menu item with the given label.

- `JCheckBoxMenuItem(String label, boolean state)`

  constructs the checkbox menu item with the given label and the given initial state (`true` is checked).

---

javax.swing.JRadioButtonMenuItem 1.2

- JRadioButtonMenuItem(String label)

  constructs the radio button menu item with the given label.

- JRadioButtonMenuItem(String label, boolean state)

  constructs the radio button menu item with the given label and the given initial state (true is checked).

---

javax.swing.AbstractButton 1.2

- boolean isSelected()
- void setSelected(boolean state)

  gets or sets the selection state of this item (true is checked).

11.5.4 Pop-Up Menus

A *pop-up menu* is a menu that is not attached to a menu bar but floats somewhere (see Figure 11.18).

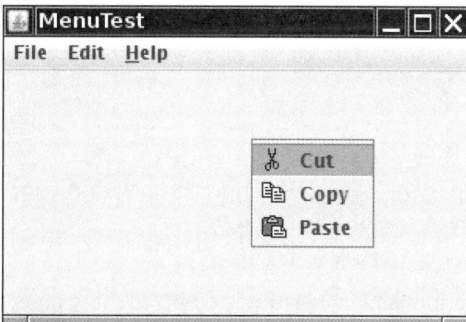

Figure 11.18 A pop-up menu

Create a pop-up menu just as you create a regular menu, except that a pop-up menu has no title.

```
var popup = new JPopupMenu();
```

Then, add your menu items as usual:

```
var item = new JMenuItem("Cut");
item.addActionListener(listener);
popup.add(item);
```

Unlike the regular menu bar that is always shown at the top of the frame, you must explicitly display a pop-up menu by using the show method. Specify the parent component and the location of the pop-up, using the coordinate system of the parent. For example:

```
popup.show(panel, x, y);
```

Usually, you want to pop up a menu when the user clicks a particular mouse button—the so-called *pop-up trigger*. In Windows and Linux, the pop-up trigger is the nonprimary (usually, the right) mouse button. To pop up a menu when the user clicks on a component, using the pop-up trigger, simply call the method

```
component.setComponentPopupMenu(popup);
```

Very occasionally, you may place a component inside another component that has a pop-up menu. The child component can inherit the parent component's pop-up menu by calling

```
child.setInheritsPopupMenu(true);
```

---

javax.swing.JPopupMenu 1.2

- void show(Component c, int x, int y)

  shows the pop-up menu over the component c with the top left corner at (x, y) (in the coordinate space of c).

- boolean isPopupTrigger(MouseEvent event) 1.3

  returns true if the mouse event is the pop-up menu trigger.

---

java.awt.event.MouseEvent 1.1

- boolean isPopupTrigger()

  returns true if this mouse event is the pop-up menu trigger.

---

javax.swing.JComponent 1.2

- JPopupMenu getComponentPopupMenu() 5
- void setComponentPopupMenu(JPopupMenu popup) 5

  gets or sets the pop-up menu for this component.

- boolean getInheritsPopupMenu() 5
- void setInheritsPopupMenu(boolean b) 5

  gets or sets the inheritsPopupMenu property. If the property is set and this component's pop-up menu is null, it uses its parent's pop-up menu.

11.5.5 Keyboard Mnemonics and Accelerators

It is a real convenience for the experienced user to select menu items by *keyboard mnemonics.* You can create a keyboard mnemonic for a menu item by specifying a mnemonic letter in the menu item constructor:

```
var aboutItem = new JMenuItem("About", 'A');
```

The keyboard mnemonic is displayed automatically in the menu, with the mnemonic letter underlined (see Figure 11.19). For example, in the item defined in the last example, the label will be displayed as "About" with an underlined letter 'A'. When the menu is displayed, the user just needs to press the A key, and the menu item is selected. (If the mnemonic letter is not part of the menu string, then typing it still selects the item, but the mnemonic is not displayed in the menu. Naturally, such invisible mnemonics are of dubious utility.)

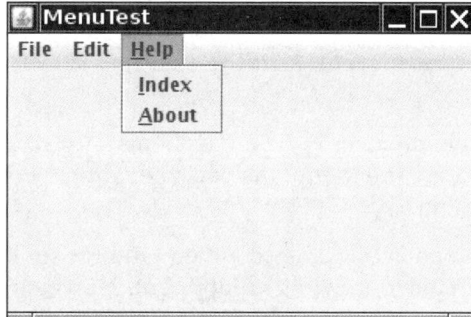

Figure 11.19 Keyboard mnemonics

Sometimes, you don't want to underline the first letter of the menu item that matches the mnemonic. For example, if you have a mnemonic 'A' for the menu item "Save As," then it makes more sense to underline the second 'A' (Save As). You can specify which character you want to have underlined by calling the setDisplayedMnemonicIndex method.

If you have an Action object, you can add the mnemonic as the value of the Action.MNEMONIC_KEY key, as follows:

```
aboutAction.putValue(Action.MNEMONIC_KEY, new Integer('A'));
```

You can supply a mnemonic letter only in the constructor of a menu item, not in the constructor for a menu. To attach a mnemonic to a menu, call the setMnemonic method:

```
var helpMenu = new JMenu("Help");
helpMenu.setMnemonic('H');
```

To select a top-level menu from the menu bar, press the Alt key together with the mnemonic letter. For example, press Alt+H to select the Help menu from the menu bar.

Keyboard mnemonics let you select a submenu or menu item from the currently open menu. In contrast, *accelerators* are keyboard shortcuts that let you select menu items without ever opening a menu. For example, many programs attach the accelerators Ctrl+O and Ctrl+S to the Open and Save items in the File menu. Use the setAccelerator method to attach an accelerator key to a menu item. The setAccelerator method takes an object of type KeyStroke. For example, the following call attaches the accelerator Ctrl+O to the openItem menu item:

```
openItem.setAccelerator(KeyStroke.getKeyStroke("ctrl O"));
```

Typing the accelerator key combination automatically selects the menu option and fires an action event, as if the user had selected the menu option manually.

You can attach accelerators only to menu items, not to menus. Accelerator keys don't actually open the menu. Instead, they directly fire the action event associated with a menu.

Conceptually, adding an accelerator to a menu item is similar to the technique of adding an accelerator to a Swing component. However, when the accelerator is added to a menu item, the key combination is automatically displayed in the menu (see Figure 11.20).

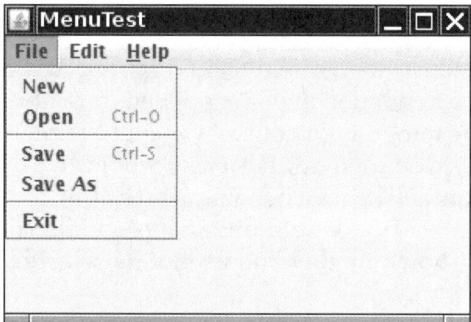

Figure 11.20 Accelerators

 NOTE: Under Windows, Alt+F4 closes a window. But this is not an accelerator to be programmed in Java. It is a shortcut defined by the operating system. This key combination will always trigger the WindowClosing event for the active window regardless of whether there is a Close item on the menu.

`javax.swing.JMenuItem` 1.2

- `JMenuItem(String label, int mnemonic)`

  constructs a menu item with a given label and mnemonic.

- `void setAccelerator(KeyStroke k)`

  sets the keystroke k as accelerator for this menu item. The accelerator key is displayed next to the label.

`javax.swing.AbstractButton` 1.2

- `void setMnemonic(int mnemonic)`

  sets the mnemonic character for the button. This character will be underlined in the label.

- `void setDisplayedMnemonicIndex(int index)` 1.4

  sets the index of the character to be underlined in the button text. Use this method if you don't want the first occurrence of the mnemonic character to be underlined.

11.5.6 Enabling and Disabling Menu Items

Occasionally, a particular menu item should be selected only in certain contexts. For example, when a document is opened in read-only mode, the Save menu item is not meaningful. Of course, we could remove the item from the menu with the `JMenu.remove` method, but users would react with some surprise to menus whose content keeps changing. Instead, it is better to deactivate the menu items that lead to temporarily inappropriate commands. A deactivated menu item is shown in gray and cannot be selected (see Figure 11.21).

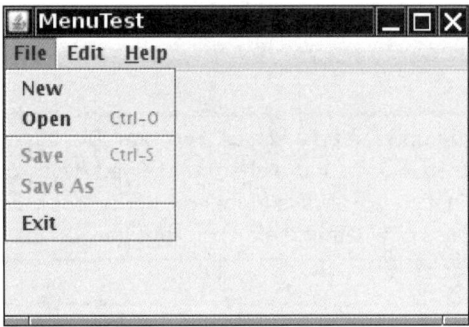

Figure 11.21 Disabled menu items

To enable or disable a menu item, use the `setEnabled` method:

```
saveItem.setEnabled(false);
```

There are two strategies for enabling and disabling menu items. Each time circumstances change, you can call `setEnabled` on the relevant menu items or actions. For example, as soon as a document has been set to read-only mode, you can locate the Save and Save As menu items and disable them. Alternatively, you can disable items just before displaying the menu. To do this, you must register a listener for the "menu selected" event. The `javax.swing.event` package defines a `MenuListener` interface with three methods:

```
void menuSelected(MenuEvent event)
void menuDeselected(MenuEvent event)
void menuCanceled(MenuEvent event)
```

The `menuSelected` method is called *before* the menu is displayed. It can therefore be used to disable or enable menu items. The following code shows how to disable the Save and Save As actions whenever the Read Only checkbox menu item is selected:

```
public void menuSelected(MenuEvent event)
{
   saveAction.setEnabled(!readonlyItem.isSelected());
   saveAsAction.setEnabled(!readonlyItem.isSelected());
}
```

 CAUTION: Disabling menu items just before displaying the menu is a clever idea, but it does not work for menu items that also have accelerator keys. Since the menu is never opened when the accelerator key is pressed, the action is never disabled, and is still triggered by the accelerator key.

`javax.swing.JMenuItem` 1.2

- void setEnabled(boolean b)

  enables or disables the menu item.

`javax.swing.event.MenuListener` 1.2

- void menuSelected(MenuEvent e)

  is called when the menu has been selected, before it is opened.

- void menuDeselected(MenuEvent e)

  is called when the menu has been deselected, after it has been closed.

- void menuCanceled(MenuEvent e)

  is called when the menu has been canceled, for example, by a user clicking outside the menu.

Listing 11.6 is a sample program that generates a set of menus. It shows all the features that you saw in this section: nested menus, disabled menu items, checkbox and radio button menu items, a pop-up menu, and keyboard mnemonics and accelerators.

Listing 11.6 menu/MenuFrame.java

```
1  package menu;
2
3  import java.awt.event.*;
4  import javax.swing.*;
5
```

(Continues)

Listing 11.6 *(Continued)*

```
6   /**
7    * A frame with a sample menu bar.
8    */
9   public class MenuFrame extends JFrame
10  {
11     private static final int DEFAULT_WIDTH = 300;
12     private static final int DEFAULT_HEIGHT = 200;
13     private Action saveAction;
14     private Action saveAsAction;
15     private JCheckBoxMenuItem readonlyItem;
16     private JPopupMenu popup;
17
18     /**
19      * A sample action that prints the action name to System.out.
20      */
21     class TestAction extends AbstractAction
22     {
23        public TestAction(String name)
24        {
25           super(name);
26        }
27
28        public void actionPerformed(ActionEvent event)
29        {
30           System.out.println(getValue(Action.NAME) + " selected.");
31        }
32     }
33
34     public MenuFrame()
35     {
36        setSize(DEFAULT_WIDTH, DEFAULT_HEIGHT);
37
38        var fileMenu = new JMenu("File");
39        fileMenu.add(new TestAction("New"));
40
41        // demonstrate accelerators
42
43        var openItem = fileMenu.add(new TestAction("Open"));
44        openItem.setAccelerator(KeyStroke.getKeyStroke("ctrl O"));
45
46        fileMenu.addSeparator();
47
48        saveAction = new TestAction("Save");
49        JMenuItem saveItem = fileMenu.add(saveAction);
50        saveItem.setAccelerator(KeyStroke.getKeyStroke("ctrl S"));
51
52        saveAsAction = new TestAction("Save As");
```

```
53      fileMenu.add(saveAsAction);
54      fileMenu.addSeparator();
55
56      fileMenu.add(new AbstractAction("Exit")
57         {
58            public void actionPerformed(ActionEvent event)
59            {
60               System.exit(0);
61            }
62         });
63
64      // demonstrate checkbox and radio button menus
65
66      readonlyItem = new JCheckBoxMenuItem("Read-only");
67      readonlyItem.addActionListener(new ActionListener()
68         {
69            public void actionPerformed(ActionEvent event)
70            {
71               boolean saveOk = !readonlyItem.isSelected();
72               saveAction.setEnabled(saveOk);
73               saveAsAction.setEnabled(saveOk);
74            }
75         });
76
77      var group = new ButtonGroup();
78
79      var insertItem = new JRadioButtonMenuItem("Insert");
80      insertItem.setSelected(true);
81      var overtypeItem = new JRadioButtonMenuItem("Overtype");
82
83      group.add(insertItem);
84      group.add(overtypeItem);
85
86      // demonstrate icons
87
88      var cutAction = new TestAction("Cut");
89      cutAction.putValue(Action.SMALL_ICON, new ImageIcon("cut.gif"));
90      var copyAction = new TestAction("Copy");
91      copyAction.putValue(Action.SMALL_ICON, new ImageIcon("copy.gif"));
92      var pasteAction = new TestAction("Paste");
93      pasteAction.putValue(Action.SMALL_ICON, new ImageIcon("paste.gif"));
94
95      var editMenu = new JMenu("Edit");
96      editMenu.add(cutAction);
97      editMenu.add(copyAction);
98      editMenu.add(pasteAction);
99
100     // demonstrate nested menus
101
```

(Continues)

Listing 11.6 *(Continued)*

```
102        var optionMenu = new JMenu("Options");
103
104        optionMenu.add(readonlyItem);
105        optionMenu.addSeparator();
106        optionMenu.add(insertItem);
107        optionMenu.add(overtypeItem);
108
109        editMenu.addSeparator();
110        editMenu.add(optionMenu);
111
112        // demonstrate mnemonics
113
114        var helpMenu = new JMenu("Help");
115        helpMenu.setMnemonic('H');
116
117        var indexItem = new JMenuItem("Index");
118        indexItem.setMnemonic('I');
119        helpMenu.add(indexItem);
120
121        // you can also add the mnemonic key to an action
122        var aboutAction = new TestAction("About");
123        aboutAction.putValue(Action.MNEMONIC_KEY, new Integer('A'));
124        helpMenu.add(aboutAction);
125
126        // add all top-level menus to menu bar
127
128        var menuBar = new JMenuBar();
129        setJMenuBar(menuBar);
130
131        menuBar.add(fileMenu);
132        menuBar.add(editMenu);
133        menuBar.add(helpMenu);
134
135        // demonstrate pop-ups
136
137        popup = new JPopupMenu();
138        popup.add(cutAction);
139        popup.add(copyAction);
140        popup.add(pasteAction);
141
142        var panel = new JPanel();
143        panel.setComponentPopupMenu(popup);
144        add(panel);
145    }
146 }
```

11.5.7 Toolbars

A toolbar is a button bar that gives quick access to the most commonly used commands in a program (see Figure 11.22).

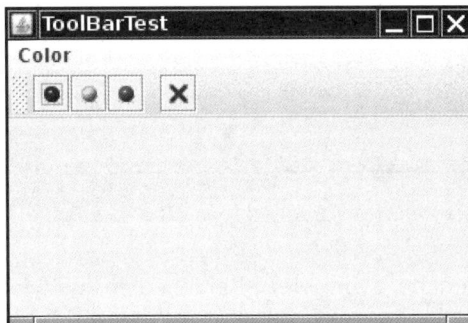

Figure 11.22 A toolbar

What makes toolbars special is that you can move them elsewhere. You can drag the toolbar to one of the four borders of the frame (see Figure 11.23). When you release the mouse button, the toolbar is dropped into the new location (see Figure 11.24).

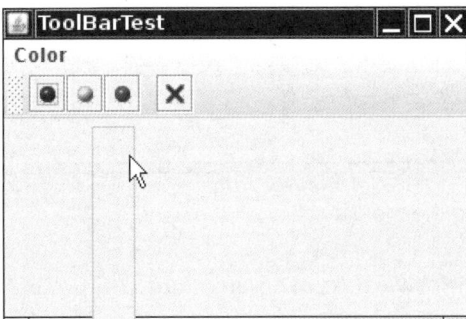

Figure 11.23 Dragging the toolbar

 NOTE: Toolbar dragging works if the toolbar is inside a container with a border layout, or any other layout manager that supports the North, East, South, and West constraints.

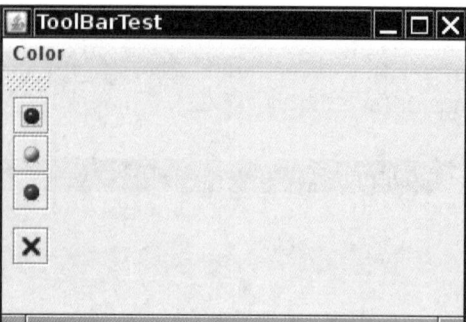

Figure 11.24 The toolbar has been dragged to another border

The toolbar can even be completely detached from the frame. A detached toolbar is contained in its own frame (see Figure 11.25). When you close the frame containing a detached toolbar, the toolbar jumps back into the original frame.

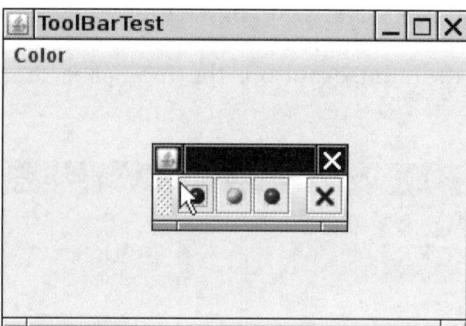

Figure 11.25 Detaching the toolbar

Toolbars are straightforward to program. Add components into the toolbar:

```
var toolbar = new JToolBar();
toolbar.add(blueButton);
```

The JToolBar class also has a method to add an Action object. Simply populate the toolbar with Action objects, like this:

```
toolbar.add(blueAction);
```

The small icon of the action is displayed in the toolbar.

You can separate groups of buttons with a separator:

```
toolbar.addSeparator();
```

For example, the toolbar in Figure 11.22 has a separator between the third and fourth button.

Then, add the toolbar to the frame:

```
add(toolbar, BorderLayout.NORTH);
```

You can also specify a title for the toolbar that appears when the toolbar is undocked:

```
toolbar = new JToolBar(titleString);
```

By default, toolbars are initially horizontal. To have a toolbar start out vertical, use

```
toolbar = new JToolBar(SwingConstants.VERTICAL)
```

or

```
toolbar = new JToolBar(titleString, SwingConstants.VERTICAL)
```

Buttons are the most common components inside toolbars. But there is no restriction on the components that you can add to a toolbar. For example, you can add a combo box to a toolbar.

11.5.8 Tooltips

A disadvantage of toolbars is that users are often mystified by the meanings of the tiny icons in toolbars. To solve this problem, user interface designers invented *tooltips*. A tooltip is activated when the cursor rests for a moment over a button. The tooltip text is displayed inside a colored rectangle. When the user moves the mouse away, the tooltip disappears. (See Figure 11.26.)

In Swing, you can add tooltips to any JComponent simply by calling the setToolTipText method:

```
exitButton.setToolTipText("Exit");
```

Alternatively, if you use Action objects, you associate the tooltip with the SHORT_DESCRIPTION key:

```
exitAction.putValue(Action.SHORT_DESCRIPTION, "Exit");
```

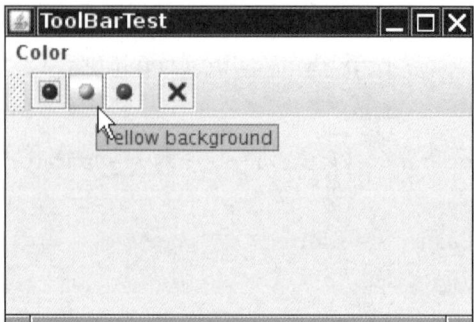

Figure 11.26 A tooltip

`javax.swing.JToolBar` 1.2

- `JToolBar()`
- `JToolBar(String titleString)`
- `JToolBar(int orientation)`
- `JToolBar(String titleString, int orientation)`

  constructs a toolbar with the given title string and orientation. `orientation` is one of `SwingConstants.HORIZONTAL` (the default) or `SwingConstants.VERTICAL`.

- `JButton add(Action a)`

  constructs a new button inside the toolbar with name, icon, short description, and action callback from the given action, and adds the button to the end of the toolbar.

- `void addSeparator()`

  adds a separator to the end of the toolbar.

`javax.swing.JComponent` 1.2

- `void setToolTipText(String text)`

  sets the text that should be displayed as a tooltip when the mouse hovers over the component.

11.6 Sophisticated Layout Management

So far we've been using only the border layout, flow layout, and grid layout for the user interface of our sample applications. For more complex tasks, this is not going to be enough.

Since Java 1.0, the AWT includes the *grid bag layout* that lays out components in rows and columns. The row and column sizes are flexible, and components can span multiple rows and columns. This layout manager is very flexible, but also very complex. The mere mention of the words "grid bag layout" has been known to strike fear in the hearts of Java programmers.

In an unsuccessful attempt to design a layout manager that would free programmers from the tyranny of the grid bag layout, the Swing designers came up with the *box layout*. According to the JDK documentation of the BoxLayout class: "Nesting multiple panels with different combinations of horizontal and vertical [*sic*] gives an effect similar to GridBagLayout, without the complexity." However, as each box is laid out independently, you cannot use box layouts to arrange neighboring components both horizontally and vertically.

Java 1.4 saw yet another attempt to design a replacement for the grid bag layout—the *spring layout* where you use imaginary springs to connect the components in a container. As the container is resized, the springs stretch or shrink, thereby adjusting the positions of the components. This sounds tedious and confusing, and it is. The spring layout quickly sank into obscurity.

The NetBeans IDE combines a layout tool (called "Matisse") and a layout manager. A user interface designer uses the tool to drop components into a container and to indicate which components should line up. The tool translates the designer's intentions into instructions for the *group layout manager*. This is much more convenient than writing the layout management code by hand.

In the coming sections, we will cover the grid bag layout because it is commonly used and is still the easiest mechanism for programmatically producing layout code. We will show you a strategy that makes grid bag layouts relatively painless in common situations.

Finally, you will see how to write your own layout manager.

11.6.1 The Grid Bag Layout

The grid bag layout is the mother of all layout managers. You can think of a grid bag layout as a grid layout without the limitations. In a grid bag layout, the rows and columns can have variable sizes. You can join adjacent cells to make room for larger components. (Many word processors, as well as HTML, provide similar capabilities for tables: You can start out with a grid and then merge adjacent cells as necessary.) The components need not fill the entire cell area, and you can specify their alignment within cells.

Consider the font selector of Figure 11.27. It consists of the following components:

- Two combo boxes to specify the font face and size
- Labels for these two combo boxes
- Two checkboxes to select bold and italic
- A text area for the sample string

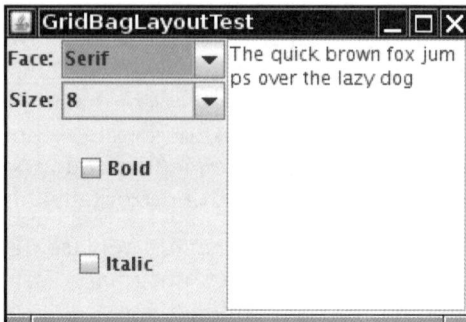

Figure 11.27 A font selector

Now, chop up the container into a grid of cells, as shown in Figure 11.28. (The rows and columns need not have equal size.) Each checkbox spans two columns, and the text area spans four rows.

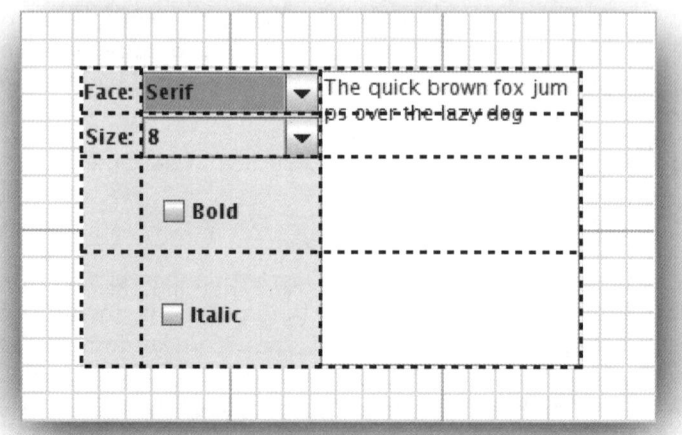

Figure 11.28 Dialog box grid used in design

To describe the layout to the grid bag manager, use the following procedure:

1. Create an object of type GridBagLayout. You don't need to tell it how many rows and columns the underlying grid has. Instead, the layout manager will try to guess it from the information you give it later.

2. Set this GridBagLayout object to be the layout manager for the component.

3. For each component, create an object of type GridBagConstraints. Set field values of the GridBagConstraints object to specify how the components are laid out within the grid bag.

4. Finally, add each component with its constraints by using the call

    ```
    add(component, constraints);
    ```

Here's an example of the code needed. (We'll go over the various constraints in more detail in the sections that follow—so don't worry if you don't know what some of the constraints do.)

```
var layout = new GridBagLayout();
panel.setLayout(layout);
var constraints = new GridBagConstraints();
constraints.weightx = 100;
constraints.weighty = 100;
constraints.gridx = 0;
constraints.gridy = 2;
constraints.gridwidth = 2;
constraints.gridheight = 1;
panel.add(component, constraints);
```

The trick is knowing how to set the state of the GridBagConstraints object. We'll discuss this object in the sections that follow.

11.6.1.1 The gridx, gridy, gridwidth, and gridheight Parameters

The gridx, gridy, gridwidth, and gridheight constraints define where the component is located in the grid. The gridx and gridy values specify the column and row positions of the upper left corner of the component to be added. The gridwidth and gridheight values determine how many columns and rows the component occupies.

The grid coordinates start with 0. In particular, gridx = 0 and gridy = 0 denotes the top left corner. The text area in our example has gridx = 2, gridy = 0 because it starts in column 2 (that is, the third column) of row 0. It has gridwidth = 1 and gridheight = 4 because it spans one column and four rows.

11.6.1.2 Weight Fields

You always need to set the *weight* fields (weightx and weighty) for each area in a grid bag layout. If you set the weight to 0, the area never grows or shrinks beyond its initial size in that direction. In the grid bag layout for Figure 11.27, we set the weightx field of the labels to be 0. This allows the labels to keep constant width when you resize the window. On the other hand, if you set the weights for all areas to 0, the container will huddle in the center of its allotted area instead of stretching to fill it.

Conceptually, the problem with the weight parameters is that weights are properties of rows and columns, not individual cells. But you need to specify them for cells because the grid bag layout does not expose the rows and columns. The row and column weights are computed as the maxima of the cell weights in each row or column. Thus, if you want a row or column to stay at a fixed size, you need to set the weights of all components in it to zero.

Note that the weights don't actually give the relative sizes of the columns. They tell what proportion of the "slack" space should be allocated to each area if the container exceeds its preferred size. This isn't particularly intuitive. We recommend that you set all weights at 100. Then, run the program and see how the layout looks. Resize the dialog to see how the rows and columns adjust. If you find that a particular row or column should not grow, set the weights of all components in it to zero. You can tinker with other weight values, but it is usually not worth the effort.

11.6.1.3 The fill and anchor Parameters

If you don't want a component to stretch out and fill the entire area, set the fill constraint. You have four possibilities for this parameter: the valid values are GridBagConstraints.NONE, GridBagConstraints.HORIZONTAL, GridBagConstraints.VERTICAL, and GridBagConstraints.BOTH.

If the component does not fill the entire area, you can specify where in the area you want it by setting the anchor field. The valid values are GridBagConstraints.CENTER (the default), GridBagConstraints.NORTH, GridBagConstraints.NORTHEAST, GridBagConstraints.EAST, and so on.

11.6.1.4 Padding

You can surround a component with additional blank space by setting the insets field of GridBagConstraints. Set the left, top, right, and bottom values of the Insets object to the amount of space that you want to have around the component. This is called the *external padding*.

The ipadx and ipady values set the *internal padding*. These values are added to the minimum width and height of the component. This ensures that the component does not shrink down to its minimum size.

11.6.1.5 Alternative Method to Specify the gridx, gridy, gridwidth, and gridheight Parameters

The AWT documentation recommends that instead of setting the gridx and gridy values to absolute positions, you set them to the constant GridBagConstraints .RELATIVE. Then, add the components to the grid bag layout in a standardized order, going from left to right in the first row, then moving along the next row, and so on.

You would still specify the number of rows and columns spanned, by giving the appropriate gridheight and gridwidth fields. However, if the component extends to the *last* row or column, you don't need to specify the actual number, but the constant GridBagConstraints.REMAINDER. This tells the layout manager that the component is the last one in its row.

This scheme does seem to work. But it sounds really goofy to hide the actual placement information from the layout manager and hope that it will rediscover it.

11.6.1.6 A Grid Bag Layout Recipe

In practice, the following recipe makes grid bag layouts relatively trouble-free:

1. Sketch out the component layout on a piece of paper.
2. Find a grid such that the small components are each contained in a cell and the larger components span multiple cells.
3. Label the rows and columns of your grid with 0, 1, 2, 3, . . . You can now read off the gridx, gridy, gridwidth, and gridheight values.
4. For each component, ask yourself whether it needs to fill its cell horizontally or vertically. If not, how do you want it aligned? This tells you the fill and anchor parameters.
5. Set all weights to 100. However, if you want a particular row or column to always stay at its default size, set the weightx or weighty to 0 in all components that belong to that row or column.
6. Write the code. Carefully double-check your settings for the GridBagConstraints. One wrong constraint can ruin your whole layout.
7. Compile, run, and enjoy.

11.6.1.7 A Helper Class to Tame the Grid Bag Constraints

The most tedious aspect of the grid bag layout is writing the code that sets the constraints. Most programmers write helper functions or a small helper class for this purpose. We present such a class after the complete code for the font dialog example. This class has the following features:

- Its name is short: GBC instead of GridBagConstraints.

- It extends GridBagConstraints, so you can use shorter names such as GBC.EAST for the constants.

- Use a GBC object when adding a component, such as

    ```
    add(component, new GBC(1, 2));
    ```

- There are two constructors to set the most common parameters: gridx and gridy, or gridx, gridy, gridwidth, and gridheight.

    ```
    add(component, new GBC(1, 2, 1, 4));
    ```

- There are convenient setters for the fields that come in x/y pairs:

    ```
    add(component, new GBC(1, 2).setWeight(100, 100));
    ```

- The setter methods return this, so you can chain them:

    ```
    add(component, new GBC(1, 2).setAnchor(GBC.EAST).setWeight(100, 100));
    ```

- The setInsets methods construct the Insets object for you. To get one-pixel insets, simply call

    ```
    add(component, new GBC(1, 2).setAnchor(GBC.EAST).setInsets(1));
    ```

Listing 11.7 shows the frame class for the font dialog example. The GBC helper class is in Listing 11.8. Here is the code that adds the components to the grid bag:

```
add(faceLabel, new GBC(0, 0).setAnchor(GBC.EAST));
add(face, new GBC(1, 0).setFill(GBC.HORIZONTAL).setWeight(100, 0).setInsets(1));
add(sizeLabel, new GBC(0, 1).setAnchor(GBC.EAST));
add(size, new GBC(1, 1).setFill(GBC.HORIZONTAL).setWeight(100, 0).setInsets(1));
add(bold, new GBC(0, 2, 2, 1).setAnchor(GBC.CENTER).setWeight(100, 100));
add(italic, new GBC(0, 3, 2, 1).setAnchor(GBC.CENTER).setWeight(100, 100));
add(sample, new GBC(2, 0, 1, 4).setFill(GBC.BOTH).setWeight(100, 100));
```

Once you understand the grid bag constraints, this kind of code is fairly easy to read and debug.

Listing 11.7 gridbag/FontFrame.java

```
1  package gridbag;
2
3  import java.awt.Font;
4  import java.awt.GridBagLayout;
5  import java.awt.event.ActionListener;
6
7  import javax.swing.BorderFactory;
8  import javax.swing.JCheckBox;
9  import javax.swing.JComboBox;
10 import javax.swing.JFrame;
11 import javax.swing.JLabel;
12 import javax.swing.JTextArea;
13
14 /**
15  * A frame that uses a grid bag layout to arrange font selection components.
16  */
17 public class FontFrame extends JFrame
18 {
19    public static final int TEXT_ROWS = 10;
20    public static final int TEXT_COLUMNS = 20;
21
22    private JComboBox<String> face;
23    private JComboBox<Integer> size;
24    private JCheckBox bold;
25    private JCheckBox italic;
26    private JTextArea sample;
27
28    public FontFrame()
29    {
30       var layout = new GridBagLayout();
31       setLayout(layout);
32
33       ActionListener listener = event -> updateSample();
34
35       // construct components
36
37       var faceLabel = new JLabel("Face: ");
38
39       face = new JComboBox<>(new String[] { "Serif", "SansSerif", "Monospaced",
40          "Dialog", "DialogInput" });
41
42       face.addActionListener(listener);
43
44       var sizeLabel = new JLabel("Size: ");
45
46       size = new JComboBox<>(new Integer[] { 8, 10, 12, 15, 18, 24, 36, 48 });
47
```

(Continues)

Listing 11.7 *(Continued)*

```
48      size.addActionListener(listener);
49
50      bold = new JCheckBox("Bold");
51      bold.addActionListener(listener);
52
53      italic = new JCheckBox("Italic");
54      italic.addActionListener(listener);
55
56      sample = new JTextArea(TEXT_ROWS, TEXT_COLUMNS);
57      sample.setText("The quick brown fox jumps over the lazy dog");
58      sample.setEditable(false);
59      sample.setLineWrap(true);
60      sample.setBorder(BorderFactory.createEtchedBorder());
61
62      // add components to grid, using GBC convenience class
63
64      add(faceLabel, new GBC(0, 0).setAnchor(GBC.EAST));
65      add(face, new GBC(1, 0).setFill(GBC.HORIZONTAL).setWeight(100, 0).setInsets(1));
66      add(sizeLabel, new GBC(0, 1).setAnchor(GBC.EAST));
67      add(size, new GBC(1, 1).setFill(GBC.HORIZONTAL).setWeight(100, 0).setInsets(1));
68      add(bold, new GBC(0, 2, 2, 1).setAnchor(GBC.CENTER).setWeight(100, 100));
69      add(italic, new GBC(0, 3, 2, 1).setAnchor(GBC.CENTER).setWeight(100, 100));
70      add(sample, new GBC(2, 0, 1, 4).setFill(GBC.BOTH).setWeight(100, 100));
71      pack();
72      updateSample();
73   }
74
75   public void updateSample()
76   {
77      var fontFace = (String) face.getSelectedItem();
78      int fontStyle = (bold.isSelected() ? Font.BOLD : 0)
79         + (italic.isSelected() ? Font.ITALIC : 0);
80      int fontSize = size.getItemAt(size.getSelectedIndex());
81      var font = new Font(fontFace, fontStyle, fontSize);
82      sample.setFont(font);
83      sample.repaint();
84   }
85 }
```

Listing 11.8 gridbag/GBC.java

```
1 package gridbag;
2
3 import java.awt.*;
4
5 /**
6  * This class simplifies the use of the GridBagConstraints class.
```

```
 7    * @version 1.01 2004-05-06
 8    * @author Cay Horstmann
 9    */
10   public class GBC extends GridBagConstraints
11   {
12      /**
13       * Constructs a GBC with a given gridx and gridy position and all other grid
14       * bag constraint values set to the default.
15       * @param gridx the gridx position
16       * @param gridy the gridy position
17       */
18      public GBC(int gridx, int gridy)
19      {
20         this.gridx = gridx;
21         this.gridy = gridy;
22      }
23
24      /**
25       * Constructs a GBC with given gridx, gridy, gridwidth, gridheight and all
26       * other grid bag constraint values set to the default.
27       * @param gridx the gridx position
28       * @param gridy the gridy position
29       * @param gridwidth the cell span in x-direction
30       * @param gridheight the cell span in y-direction
31       */
32      public GBC(int gridx, int gridy, int gridwidth, int gridheight)
33      {
34         this.gridx = gridx;
35         this.gridy = gridy;
36         this.gridwidth = gridwidth;
37         this.gridheight = gridheight;
38      }
39
40      /**
41       * Sets the anchor.
42       * @param anchor the anchor value
43       * @return this object for further modification
44       */
45      public GBC setAnchor(int anchor)
46      {
47         this.anchor = anchor;
48         return this;
49      }
50
51      /**
52       * Sets the fill direction.
53       * @param fill the fill direction
54       * @return this object for further modification
55       */
```

(Continues)

Listing 11.8 *(Continued)*

```
56      public GBC setFill(int fill)
57      {
58         this.fill = fill;
59         return this;
60      }
61
62      /**
63       * Sets the cell weights.
64       * @param weightx the cell weight in x-direction
65       * @param weighty the cell weight in y-direction
66       * @return this object for further modification
67       */
68      public GBC setWeight(double weightx, double weighty)
69      {
70         this.weightx = weightx;
71         this.weighty = weighty;
72         return this;
73      }
74
75      /**
76       * Sets the insets of this cell.
77       * @param distance the spacing to use in all directions
78       * @return this object for further modification
79       */
80      public GBC setInsets(int distance)
81      {
82         this.insets = new Insets(distance, distance, distance, distance);
83         return this;
84      }
85
86      /**
87       * Sets the insets of this cell.
88       * @param top the spacing to use on top
89       * @param left the spacing to use to the left
90       * @param bottom the spacing to use on the bottom
91       * @param right the spacing to use to the right
92       * @return this object for further modification
93       */
94      public GBC setInsets(int top, int left, int bottom, int right)
95      {
96         this.insets = new Insets(top, left, bottom, right);
97         return this;
98      }
99
```

```
100    /**
101     * Sets the internal padding
102     * @param ipadx the internal padding in x-direction
103     * @param ipady the internal padding in y-direction
104     * @return this object for further modification
105     */
106    public GBC setIpad(int ipadx, int ipady)
107    {
108       this.ipadx = ipadx;
109       this.ipady = ipady;
110       return this;
111    }
112 }
```

---

java.awt.GridBagConstraints 1.0

- int gridx, gridy

  specifies the starting column and row of the cell. The default is 0.

- int gridwidth, gridheight

  specifies the column and row extent of the cell. The default is 1.

- double weightx, weighty

  specifies the capacity of the cell to grow. The default is 0.

- int anchor

  indicates the alignment of the component inside the cell. You can choose between absolute positions:

NORTHWEST	NORTH	NORTHEAST
WEST	CENTER	EAST
SOUTHWEST	SOUTH	SOUTHEAST

  or their orientation-independent counterparts:

FIRST_LINE_START	LINE_START	FIRST_LINE_END
PAGE_START	CENTER	PAGE_END
LAST_LINE_START	LINE_END	LAST_LINE_END

  Use the latter if your application may be localized for right-to-left or top-to-bottom text. The default is CENTER.

- int fill

  specifies the fill behavior of the component inside the cell: one of NONE, BOTH, HORIZONTAL, or VERTICAL. The default is NONE.

(Continues)

---

java.awt.GridBagConstraints 1.0   *(Continued)*

- int ipadx, ipady

  specifies the "internal" padding around the component. The default is 0.

- Insets insets

  specifies the "external" padding along the cell boundaries. The default is no padding.

- GridBagConstraints(int gridx, int gridy, int gridwidth, int gridheight, double weightx, double weighty, int anchor, int fill, Insets insets, int ipadx, int ipady) 1.2

  constructs a GridBagConstraints with all its fields specified in the arguments. This constructor should only be used by automatic code generators because it makes your source code very hard to read.

11.6.2 Custom Layout Managers

You can design your own LayoutManager class that manages components in a special way. As a fun example, let's arrange all components in a container to form a circle (see Figure 11.29).

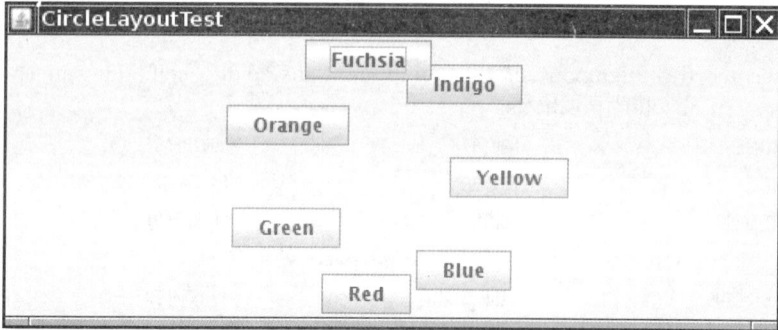

Figure 11.29 Circle layout

Your own layout manager must implement the LayoutManager interface. You need to override the following five methods:

```
void addLayoutComponent(String s, Component c)
void removeLayoutComponent(Component c)
Dimension preferredLayoutSize(Container parent)
Dimension minimumLayoutSize(Container parent)
void layoutContainer(Container parent)
```

The first two methods are called when a component is added or removed. If you don't keep any additional information about the components, you can make them do nothing. The next two methods compute the space required for the minimum and the preferred layout of the components. These are usually the same quantity. The fifth method does the actual work and invokes setBounds on all components.

> **NOTE:** The AWT has a second interface, called LayoutManager2, with ten methods to implement rather than five. The main point of the LayoutManager2 interface is to allow you to use the add method with constraints. For example, the BorderLayout and GridBagLayout implement the LayoutManager2 interface.

Listing 11.9 shows the code for the CircleLayout manager which, uselessly enough, lays out the components along a circle inside the parent. The frame class of the sample program is in Listing 11.10.

Listing 11.9 circleLayout/CircleLayout.java

```
1  package circleLayout;
2
3  import java.awt.*;
4
5  /**
6   * A layout manager that lays out components along a circle.
7   */
8  public class CircleLayout implements LayoutManager
9  {
10     private int minWidth = 0;
11     private int minHeight = 0;
12     private int preferredWidth = 0;
13     private int preferredHeight = 0;
14     private boolean sizesSet = false;
15     private int maxComponentWidth = 0;
16     private int maxComponentHeight = 0;
17
18     public void addLayoutComponent(String name, Component comp)
19     {
20     }
21
22     public void removeLayoutComponent(Component comp)
23     {
24     }
25
```

(Continues)

Listing 11.9 *(Continued)*

```
26    public void setSizes(Container parent)
27    {
28       if (sizesSet) return;
29       int n = parent.getComponentCount();
30
31       preferredWidth = 0;
32       preferredHeight = 0;
33       minWidth = 0;
34       minHeight = 0;
35       maxComponentWidth = 0;
36       maxComponentHeight = 0;
37
38       // compute the maximum component widths and heights
39       // and set the preferred size to the sum of the component sizes
40       for (int i = 0; i < n; i++)
41       {
42          Component c = parent.getComponent(i);
43          if (c.isVisible())
44          {
45             Dimension d = c.getPreferredSize();
46             maxComponentWidth = Math.max(maxComponentWidth, d.width);
47             maxComponentHeight = Math.max(maxComponentHeight, d.height);
48             preferredWidth += d.width;
49             preferredHeight += d.height;
50          }
51       }
52       minWidth = preferredWidth / 2;
53       minHeight = preferredHeight / 2;
54       sizesSet = true;
55    }
56
57    public Dimension preferredLayoutSize(Container parent)
58    {
59       setSizes(parent);
60       Insets insets = parent.getInsets();
61       int width = preferredWidth + insets.left + insets.right;
62       int height = preferredHeight + insets.top + insets.bottom;
63       return new Dimension(width, height);
64    }
65
66    public Dimension minimumLayoutSize(Container parent)
67    {
68       setSizes(parent);
69       Insets insets = parent.getInsets();
```

```
70      int width = minWidth + insets.left + insets.right;
71      int height = minHeight + insets.top + insets.bottom;
72      return new Dimension(width, height);
73   }
74
75   public void layoutContainer(Container parent)
76   {
77      setSizes(parent);
78
79      // compute center of the circle
80
81      Insets insets = parent.getInsets();
82      int containerWidth = parent.getSize().width - insets.left - insets.right;
83      int containerHeight = parent.getSize().height - insets.top - insets.bottom;
84
85      int xcenter = insets.left + containerWidth / 2;
86      int ycenter = insets.top + containerHeight / 2;
87
88      // compute radius of the circle
89
90      int xradius = (containerWidth - maxComponentWidth) / 2;
91      int yradius = (containerHeight - maxComponentHeight) / 2;
92      int radius = Math.min(xradius, yradius);
93
94      // lay out components along the circle
95
96      int n = parent.getComponentCount();
97      for (int i = 0; i < n; i++)
98      {
99         Component c = parent.getComponent(i);
100        if (c.isVisible())
101        {
102           double angle = 2 * Math.PI * i / n;
103
104           // center point of component
105           int x = xcenter + (int) (Math.cos(angle) * radius);
106           int y = ycenter + (int) (Math.sin(angle) * radius);
107
108           // move component so that its center is (x, y)
109           // and its size is its preferred size
110           Dimension d = c.getPreferredSize();
111           c.setBounds(x - d.width / 2, y - d.height / 2, d.width, d.height);
112        }
113     }
114  }
115 }
```

Listing 11.10 circleLayout/CircleLayoutFrame.java

```
1 package circleLayout;
2
3 import javax.swing.*;
4
5 /**
6  * A frame that shows buttons arranged along a circle.
7  */
8 public class CircleLayoutFrame extends JFrame
9 {
10    public CircleLayoutFrame()
11    {
12       setLayout(new CircleLayout());
13       add(new JButton("Yellow"));
14       add(new JButton("Blue"));
15       add(new JButton("Red"));
16       add(new JButton("Green"));
17       add(new JButton("Orange"));
18       add(new JButton("Fuchsia"));
19       add(new JButton("Indigo"));
20       pack();
21    }
22 }
```

---

java.awt.LayoutManager 1.0

- void addLayoutComponent(String name, Component comp)

  adds a component to the layout.

- void removeLayoutComponent(Component comp)

  removes a component from the layout.

- Dimension preferredLayoutSize(Container cont)

  returns the preferred size dimensions for the container under this layout.

- Dimension minimumLayoutSize(Container cont)

  returns the minimum size dimensions for the container under this layout.

- void layoutContainer(Container cont)

  lays out the components in a container.

11.7 Dialog Boxes

So far, all our user interface components have appeared inside a frame window that was created in the application. This is the most common situation if you write *applets* that run inside a web browser. But if you write applications, you usually want separate dialog boxes to pop up to give information to, or get information from, the user.

Just as with most windowing systems, AWT distinguishes between *modal* and *modeless* dialog boxes. A modal dialog box won't let users interact with the remaining windows of the application until he or she deals with it. Use a modal dialog box when you need information from the user before you can proceed with execution. For example, when the user wants to read a file, a modal file dialog box is the one to pop up. The user must specify a file name before the program can begin the read operation. Only when the user closes the modal dialog box can the application proceed.

A modeless dialog box lets the user enter information in both the dialog box and the remainder of the application. One example of a modeless dialog is a toolbar. The toolbar can stay in place as long as needed, and the user can interact with both the application window and the toolbar as needed.

We will start this section with the simplest dialogs—modal dialogs with just a single message. Swing has a convenient JOptionPane class that lets you put up a simple dialog without writing any special dialog box code. Next, you will see how to write more complex dialogs by implementing your own dialog windows. Finally, you will see how to transfer data from your application into a dialog and back.

We'll conclude the discussion of dialog boxes by looking at the Swing JFileChooser.

11.7.1 Option Dialogs

Swing has a set of ready-made simple dialogs that suffice to ask the user for a single piece of information. The JOptionPane has four static methods to show these simple dialogs:

showMessageDialog	Show a message and wait for the user to click OK
showConfirmDialog	Show a message and get a confirmation (like OK/Cancel)
showOptionDialog	Show a message and get a user option from a set of options
showInputDialog	Show a message and get one line of user input

Figure 11.30 shows a typical dialog. As you can see, the dialog has the following components:

- An icon
- A message
- One or more option buttons

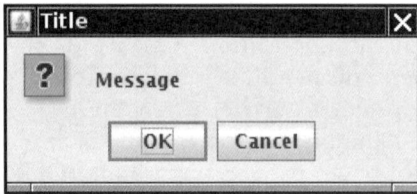

Figure 11.30 An option dialog

The input dialog has an additional component for user input. This can be a text field into which the user can type an arbitrary string, or a combo box from which the user can select one item.

The exact layout of these dialogs and the choice of icons for standard message types depend on the pluggable look-and-feel.

The icon on the left side depends on one of five *message types*:

```
ERROR_MESSAGE
INFORMATION_MESSAGE
WARNING_MESSAGE
QUESTION_MESSAGE
PLAIN_MESSAGE
```

The PLAIN_MESSAGE type has no icon. Each dialog type also has a method that lets you supply your own icon instead.

For each dialog type, you can specify a message. This message can be a string, an icon, a user interface component, or any other object. Here is how the message object is displayed:

String	Draw the string
Icon	Show the icon
Component	Show the component
Object[]	Show all objects in the array, stacked on top of each other
Any other object	Apply toString and show the resulting string

Of course, supplying a message string is by far the most common case. Supplying a Component gives you ultimate flexibility because you can make the paintComponent method draw anything you want.

The buttons at the bottom depend on the dialog type and the *option type*. When calling showMessageDialog and showInputDialog, you get only a standard set of buttons (OK and OK/Cancel, respectively). When calling showConfirmDialog, you can choose among four option types:

```
DEFAULT_OPTION
YES_NO_OPTION
YES_NO_CANCEL_OPTION
OK_CANCEL_OPTION
```

With the showOptionDialog you can specify an arbitrary set of options. You supply an array of objects for the options. Each array element is rendered as follows:

String	Make a button with the string as label
Icon	Make a button with the icon as label
Component	Show the component
Any other object	Apply toString and make a button with the resulting string as label

The return values of these functions are as follows:

showMessageDialog	None
showConfirmDialog	An integer representing the chosen option
showOptionDialog	An integer representing the chosen option
showInputDialog	The string that the user supplied or selected

The showConfirmDialog and showOptionDialog return integers to indicate which button the user chose. For the option dialog, this is simply the index of the chosen option or the value CLOSED_OPTION if the user closed the dialog instead of choosing an option. For the confirmation dialog, the return value can be one of the following:

```
OK_OPTION
CANCEL_OPTION
YES_OPTION
NO_OPTION
CLOSED_OPTION
```

This all sounds like a bewildering set of choices, but in practice it is simple. Follow these steps:

1. Choose the dialog type (message, confirmation, option, or input).
2. Choose the icon (error, information, warning, question, none, or custom).
3. Choose the message (string, icon, custom component, or a stack of them).
4. For a confirmation dialog, choose the option type (default, Yes/No, Yes/No/Cancel, or OK/Cancel).

5. For an option dialog, choose the options (strings, icons, or custom components) and the default option.

6. For an input dialog, choose between a text field and a combo box.

7. Locate the appropriate method to call in the JOptionPane API.

For example, suppose you want to show the dialog in Figure 11.30. The dialog shows a message and asks the user to confirm or cancel. Thus, it is a confirmation dialog. The icon is a question icon. The message is a string. The option type is OK_CANCEL_OPTION. Here is the call you would make:

```
int selection = JOptionPane.showConfirmDialog(parent,
    "Message", "Title",
    JOptionPane.OK_CANCEL_OPTION,
    JOptionPane.QUESTION_MESSAGE);
if (selection == JOptionPane.OK_OPTION) . . .
```

 TIP: The message string can contain newline ('\n') characters. Such a string is displayed in multiple lines.

---

javax.swing.JOptionPane 1.2

- static void showMessageDialog(Component parent, Object message, String title, int messageType, Icon icon)
- static void showMessageDialog(Component parent, Object message, String title, int messageType)
- static void showMessageDialog(Component parent, Object message)
- static void showInternalMessageDialog(Component parent, Object message, String title, int messageType, Icon icon)
- static void showInternalMessageDialog(Component parent, Object message, String title, int messageType)
- static void showInternalMessageDialog(Component parent, Object message)

  shows a message dialog or an internal message dialog. (An internal dialog is rendered entirely within its owner's frame.) The parent component can be null. The message to show on the dialog can be a string, icon, component, or an array of them. The messageType parameter is one of ERROR_MESSAGE, INFORMATION_MESSAGE, WARNING_MESSAGE, QUESTION_MESSAGE, PLAIN_MESSAGE.

(Continues)

---

javax.swing.JOptionPane `1.2` *(Continued)*

- `static int showConfirmDialog(Component parent, Object message, String title, int optionType,` `int messageType, Icon icon)`
- `static int showConfirmDialog(Component parent, Object message, String title, int optionType,` `int messageType)`
- `static int showConfirmDialog(Component parent, Object message, String title, int optionType)`
- `static int showConfirmDialog(Component parent, Object message)`
- `static int showInternalConfirmDialog(Component parent, Object message, String title, int` `optionType, int messageType, Icon icon)`
- `static int showInternalConfirmDialog(Component parent, Object message, String title, int` `optionType, int messageType)`
- `static int showInternalConfirmDialog(Component parent, Object message, String title, int` `optionType)`
- `static int showInternalConfirmDialog(Component parent, Object message)`

  shows a confirmation dialog or an internal confirmation dialog. (An internal dialog is rendered entirely within its owner's frame.) Returns the option selected by the user (one of `OK_OPTION`, `CANCEL_OPTION`, `YES_OPTION`, `NO_OPTION`), or `CLOSED_OPTION` if the user closed the dialog. The parent component can be `null`. The message to show on the dialog can be a string, icon, component, or an array of them. The `messageType` parameter is one of `ERROR_MESSAGE`, `INFORMATION_MESSAGE`, `WARNING_MESSAGE`, `QUESTION_MESSAGE`, `PLAIN_MESSAGE`, and `optionType` is one of `DEFAULT_OPTION`, `YES_NO_OPTION`, `YES_NO_CANCEL_OPTION`, `OK_CANCEL_OPTION`.

- `static int showOptionDialog(Component parent, Object message, String title, int optionType,` `int messageType, Icon icon, Object[] options, Object default)`
- `static int showInternalOptionDialog(Component parent, Object message, String title, int` `optionType, int messageType, Icon icon, Object[] options, Object default)`

  shows an option dialog or an internal option dialog. (An internal dialog is rendered entirely within its owner's frame.) Returns the index of the option selected by the user, or `CLOSED_OPTION` if the user canceled the dialog. The parent component can be `null`. The message to show on the dialog can be a string, icon, component, or an array of them. The `messageType` parameter is one of `ERROR_MESSAGE`, `INFORMATION_MESSAGE`, `WARNING_MESSAGE`, `QUESTION_MESSAGE`, `PLAIN_MESSAGE`, and `optionType` is one of `DEFAULT_OPTION`, `YES_NO_OPTION`, `YES_NO_CANCEL_OPTION`, `OK_CANCEL_OPTION`. The options parameter is an array of strings, icons, or components.

(Continues)

---

javax.swing.JOptionPane 1.2 *(Continued)*

- static Object showInputDialog(Component parent, Object message, String title, int messageType, Icon icon, Object[] values, Object default)
- static String showInputDialog(Component parent, Object message, String title, int messageType)
- static String showInputDialog(Component parent, Object message)
- static String showInputDialog(Object message)
- static String showInputDialog(Component parent, Object message, Object default) 1.4
- static String showInputDialog(Object message, Object default) 1.4
- static Object showInternalInputDialog(Component parent, Object message, String title, int messageType, Icon icon, Object[] values, Object default)
- static String showInternalInputDialog(Component parent, Object message, String title, int messageType)
- static String showInternalInputDialog(Component parent, Object message)

  shows an input dialog or an internal input dialog. (An internal dialog is rendered entirely within its owner's frame.) Returns the input string typed by the user, or null if the user canceled the dialog. The parent component can be null. The message to show on the dialog can be a string, icon, component, or an array of them. The messageType parameter is one of ERROR_MESSAGE, INFORMATION_MESSAGE, WARNING_MESSAGE, QUESTION_MESSAGE, PLAIN_MESSAGE.

11.7.2 Creating Dialogs

In the last section, you saw how to use the JOptionPane class to show a simple dialog. In this section, you will see how to create such a dialog by hand.

Figure 11.31 shows a typical modal dialog box—a program information box that is displayed when the user clicks the About button.

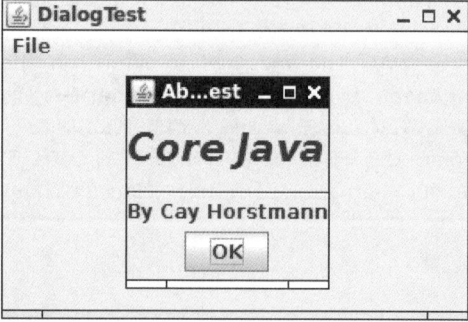

Figure 11.31 An About dialog box

To implement a dialog box, you extend the JDialog class. This is essentially the same process as extending JFrame for the main window for an application. More precisely:

1. In the constructor of your dialog box, call the constructor of the superclass JDialog.
2. Add the user interface components of the dialog box.
3. Add the event handlers.
4. Set the size for the dialog box.

When you call the superclass constructor, you will need to supply the *owner frame*, the title of the dialog, and the *modality*.

The owner frame controls where the dialog is displayed. You can supply null as the owner; then, the dialog is owned by a hidden frame.

The modality specifies which other windows of your application are blocked while the dialog is displayed. A modeless dialog does not block other windows. A modal dialog blocks all other windows of the application (except for the children of the dialog). You would use a modeless dialog for a toolbox that the user can always access. On the other hand, you would use a modal dialog if you want to force the user to supply required information before continuing.

Here's the code for a dialog box:

```
public AboutDialog extends JDialog
{
   public AboutDialog(JFrame owner)
   {
      super(owner, "About DialogTest", true);
      add(new JLabel(
         "<html><h1><i>Core Java</i></h1><hr>By Cay Horstmann</html>"),
         BorderLayout.CENTER);

      var panel = new JPanel();
      var ok = new JButton("OK");

      ok.addActionListener(event -> setVisible(false));
      panel.add(ok);
      add(panel, BorderLayout.SOUTH);
      setSize(250, 150);
   }
}
```

As you can see, the constructor adds user interface elements—in this case, labels and a button. It adds a handler to the button and sets the size of the dialog.

To display the dialog box, create a new dialog object and make it visible:

```
var dialog = new AboutDialog(this);
dialog.setVisible(true);
```

Actually, in the sample code below, we create the dialog box only once, and we can reuse it whenever the user clicks the About button.

```
if (dialog == null) // first time
   dialog = new AboutDialog(this);
dialog.setVisible(true);
```

When the user clicks the OK button, the dialog box should close. This is handled in the event handler of the OK button:

```
ok.addActionListener(event -> setVisible(false));
```

When the user closes the dialog by clicking the Close button, the dialog is also hidden. Just as with a JFrame, you can override this behavior with the setDefaultCloseOperation method.

Listing 11.11 is the code for the frame class of the test program. Listing 11.12 shows the dialog class.

Listing 11.11 dialog/DialogFrame.java

```
1  package dialog;
2
3  import javax.swing.JFrame;
4  import javax.swing.JMenu;
5  import javax.swing.JMenuBar;
6  import javax.swing.JMenuItem;
7
8  /**
9   * A frame with a menu whose File->About action shows a dialog.
10  */
11 public class DialogFrame extends JFrame
12 {
13    private static final int DEFAULT_WIDTH = 300;
14    private static final int DEFAULT_HEIGHT = 200;
15    private AboutDialog dialog;
16
17    public DialogFrame()
18    {
19       setSize(DEFAULT_WIDTH, DEFAULT_HEIGHT);
20
21       // construct a File menu
22
23       var menuBar = new JMenuBar();
```

```
24      setJMenuBar(menuBar);
25      var fileMenu = new JMenu("File");
26      menuBar.add(fileMenu);
27
28      // add About and Exit menu items
29
30      // the About item shows the About dialog
31
32      var aboutItem = new JMenuItem("About");
33      aboutItem.addActionListener(event -> {
34         if (dialog == null) // first time
35            dialog = new AboutDialog(DialogFrame.this);
36         dialog.setVisible(true); // pop up dialog
37      });
38      fileMenu.add(aboutItem);
39
40      // the Exit item exits the program
41
42      var exitItem = new JMenuItem("Exit");
43      exitItem.addActionListener(event -> System.exit(0));
44      fileMenu.add(exitItem);
45   }
46 }
```

Listing 11.12 dialog/AboutDialog.java

```
1 package dialog;
2
3 import java.awt.BorderLayout;
4
5 import javax.swing.JButton;
6 import javax.swing.JDialog;
7 import javax.swing.JFrame;
8 import javax.swing.JLabel;
9 import javax.swing.JPanel;
10
11 /**
12  * A sample modal dialog that displays a message and waits for the user to click
13  * the OK button.
14  */
15 public class AboutDialog extends JDialog
16 {
17    public AboutDialog(JFrame owner)
18    {
19       super(owner, "About DialogTest", true);
20
21       // add HTML label to center
22
```

(Continues)

Listing 11.12 *(Continued)*

```
23      add(
24        new JLabel(
25          "<html><h1><i>Core Java</i></h1><hr>By Cay Horstmann</html>"),
26        BorderLayout.CENTER);
27
28      // OK button closes the dialog
29
30      var ok = new JButton("OK");
31      ok.addActionListener(event -> setVisible(false));
32
33      // add OK button to southern border
34
35      var panel = new JPanel();
36      panel.add(ok);
37      add(panel, BorderLayout.SOUTH);
38
39      pack();
40    }
41 }
```

`javax.swing.JDialog` 1.2

- `public JDialog(Frame parent, String title, boolean modal)`

  constructs a dialog. The dialog is not visible until it is explicitly shown.

11.7.3 Data Exchange

The most common reason to put up a dialog box is to get information from the user. You have already seen how easy it is to make a dialog box object: Give it initial data and call `setVisible(true)` to display the dialog box on the screen. Now let's see how to transfer data in and out of a dialog box.

Consider the dialog box in Figure 11.32 that could be used to obtain a user name and a password to connect to some online service.

Your dialog box should provide methods to set default data. For example, the `PasswordChooser` class of the example program has a method, `setUser`, to place default values into the next fields:

```
public void setUser(User u)
{
   username.setText(u.getName());
}
```

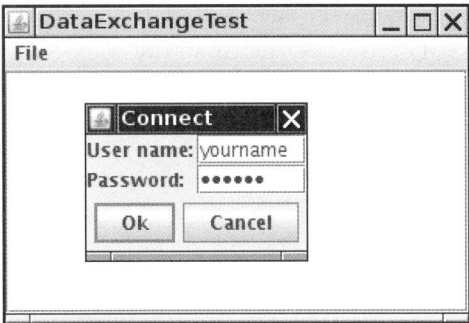

Figure 11.32 Password dialog box

Once you set the defaults (if desired), show the dialog by calling setVisible(true). The dialog is now displayed.

The user then fills in the information and clicks the OK or Cancel button. The event handlers for both buttons call setVisible(false), which terminates the call to setVisible(true). Alternatively, the user may close the dialog. If you did not install a window listener for the dialog, the default window closing operation applies: The dialog becomes invisible, which also terminates the call to setVisible(true).

The important issue is that the call to setVisible(true) blocks until the user has dismissed the dialog. This makes it easy to implement modal dialogs.

You want to know whether the user has accepted or canceled the dialog. Our sample code sets the ok flag to false before showing the dialog. Only the event handler for the OK button sets the ok flag to true; that's how you retrieve the user input from the dialog.

 NOTE: Transferring data out of a modeless dialog is not as simple. When a modeless dialog is displayed, the call to setVisible(true) does not block and the program continues running while the dialog is displayed. If the user selects items on a modeless dialog and then clicks OK, the dialog needs to send an event to some listener in the program.

The example program contains another useful improvement. When you construct a JDialog object, you need to specify the owner frame. However, quite often you want to show the same dialog with different owner frames. It is better to pick the owner frame *when you are ready to show the dialog*, not when you construct the PasswordChooser object.

The trick is to have the PasswordChooser extend JPanel instead of JDialog. Build a JDialog object on the fly in the showDialog method:

```
public boolean showDialog(Frame owner, String title)
{
   ok = false;

   if (dialog == null || dialog.getOwner() != owner)
   {
      dialog = new JDialog(owner, true);
      dialog.add(this);
      dialog.pack();
   }

   dialog.setTitle(title);
   dialog.setVisible(true);
   return ok;
}
```

Note that it is safe to have owner equal to null.

You can do even better. Sometimes, the owner frame isn't readily available. It is easy enough to compute it from any parent component, like this:

```
Frame owner;
if (parent instanceof Frame)
   owner = (Frame) parent;
else
   owner = (Frame) SwingUtilities.getAncestorOfClass(Frame.class, parent);
```

We use this enhancement in our sample program. The JOptionPane class also uses this mechanism.

Many dialogs have a *default button*, which is automatically selected if the user presses a trigger key (Enter in most look-and-feel implementations). The default button is specially marked, often with a thick outline.

Set the default button in the *root pane* of the dialog:

```
dialog.getRootPane().setDefaultButton(okButton);
```

If you follow our suggestion of laying out the dialog in a panel, then you must be careful to set the default button only after you wrapped the panel into a dialog. The panel dialog itself has no root pane.

Listing 11.13 is for the frame class of the program that illustrates the data flow into and out of a dialog box. Listing 11.14 shows the dialog class.

Listing 11.13 dataExchange/DataExchangeFrame.java

```java
 1  package dataExchange;
 2
 3  import java.awt.*;
 4  import java.awt.event.*;
 5  import javax.swing.*;
 6
 7  /**
 8   * A frame with a menu whose File->Connect action shows a password dialog.
 9   */
10  public class DataExchangeFrame extends JFrame
11  {
12     public static final int TEXT_ROWS = 20;
13     public static final int TEXT_COLUMNS = 40;
14     private PasswordChooser dialog = null;
15     private JTextArea textArea;
16
17     public DataExchangeFrame()
18     {
19        // construct a File menu
20
21        var mbar = new JMenuBar();
22        setJMenuBar(mbar);
23        var fileMenu = new JMenu("File");
24        mbar.add(fileMenu);
25
26        // add Connect and Exit menu items
27
28        var connectItem = new JMenuItem("Connect");
29        connectItem.addActionListener(new ConnectAction());
30        fileMenu.add(connectItem);
31
32        // the Exit item exits the program
33
34        var exitItem = new JMenuItem("Exit");
35        exitItem.addActionListener(event -> System.exit(0));
36        fileMenu.add(exitItem);
37
38        textArea = new JTextArea(TEXT_ROWS, TEXT_COLUMNS);
39        add(new JScrollPane(textArea), BorderLayout.CENTER);
40        pack();
41     }
42
43     /**
44      * The Connect action pops up the password dialog.
45      */
```

(Continues)

Listing 11.13 *(Continued)*

```
46    private class ConnectAction implements ActionListener
47    {
48        public void actionPerformed(ActionEvent event)
49        {
50            // if first time, construct dialog
51
52            if (dialog == null) dialog = new PasswordChooser();
53
54            // set default values
55            dialog.setUser(new User("yourname", null));
56
57            // pop up dialog
58            if (dialog.showDialog(DataExchangeFrame.this, "Connect"))
59            {
60                // if accepted, retrieve user input
61                User u = dialog.getUser();
62                textArea.append("user name = " + u.getName() + ", password = "
63                    + (new String(u.getPassword())) + "\n");
64            }
65        }
66    }
67 }
```

Listing 11.14 dataExchange/PasswordChooser.java

```
1  package dataExchange;
2
3  import java.awt.BorderLayout;
4  import java.awt.Component;
5  import java.awt.Frame;
6  import java.awt.GridLayout;
7
8  import javax.swing.JButton;
9  import javax.swing.JDialog;
10 import javax.swing.JLabel;
11 import javax.swing.JPanel;
12 import javax.swing.JPasswordField;
13 import javax.swing.JTextField;
14 import javax.swing.SwingUtilities;
15
16 /**
17  * A password chooser that is shown inside a dialog.
18  */
19 public class PasswordChooser extends JPanel
20 {
```

```
21   private JTextField username;
22   private JPasswordField password;
23   private JButton okButton;
24   private boolean ok;
25   private JDialog dialog;
26
27   public PasswordChooser()
28   {
29      setLayout(new BorderLayout());
30
31      // construct a panel with user name and password fields
32
33      var panel = new JPanel();
34      panel.setLayout(new GridLayout(2, 2));
35      panel.add(new JLabel("User name:"));
36      panel.add(username = new JTextField(""));
37      panel.add(new JLabel("Password:"));
38      panel.add(password = new JPasswordField(""));
39      add(panel, BorderLayout.CENTER);
40
41      // create Ok and Cancel buttons that terminate the dialog
42
43      okButton = new JButton("Ok");
44      okButton.addActionListener(event -> {
45         ok = true;
46         dialog.setVisible(false);
47      });
48
49      var cancelButton = new JButton("Cancel");
50      cancelButton.addActionListener(event -> dialog.setVisible(false));
51
52      // add buttons to southern border
53
54      var buttonPanel = new JPanel();
55      buttonPanel.add(okButton);
56      buttonPanel.add(cancelButton);
57      add(buttonPanel, BorderLayout.SOUTH);
58   }
59
60   /**
61    * Sets the dialog defaults.
62    * @param u the default user information
63    */
64   public void setUser(User u)
65   {
66      username.setText(u.getName());
67   }
68
```

(Continues)

Listing 11.14 *(Continued)*

```java
69     /**
70      * Gets the dialog entries.
71      * @return a User object whose state represents the dialog entries
72      */
73     public User getUser()
74     {
75        return new User(username.getText(), password.getPassword());
76     }
77
78     /**
79      * Show the chooser panel in a dialog.
80      * @param parent a component in the owner frame or null
81      * @param title the dialog window title
82      */
83     public boolean showDialog(Component parent, String title)
84     {
85        ok = false;
86
87        // locate the owner frame
88
89        Frame owner = null;
90        if (parent instanceof Frame)
91           owner = (Frame) parent;
92        else
93           owner = (Frame) SwingUtilities.getAncestorOfClass(Frame.class, parent);
94
95        // if first time, or if owner has changed, make new dialog
96
97        if (dialog == null || dialog.getOwner() != owner)
98        {
99           dialog = new JDialog(owner, true);
100          dialog.add(this);
101          dialog.getRootPane().setDefaultButton(okButton);
102          dialog.pack();
103       }
104
105       // set title and show dialog
106
107       dialog.setTitle(title);
108       dialog.setVisible(true);
109       return ok;
110    }
111 }
```

---

javax.swing.SwingUtilities 1.2

- Container getAncestorOfClass(Class c, Component comp)

  returns the innermost parent container of the given component that belongs to the given class or one of its subclasses.

---

javax.swing.JComponent 1.2

- JRootPane getRootPane()

  gets the root pane enclosing this component, or null if this component does not have an ancestor with a root pane.

---

javax.swing.JRootPane 1.2

- void setDefaultButton(JButton button)

  sets the default button for this root pane. To deactivate the default button, call this method with a null parameter.

---

javax.swing.JButton 1.2

- boolean isDefaultButton()

  returns true if this button is the default button of its root pane.

11.7.4 File Dialogs

In an application, you often want to be able to open and save files. A good file dialog box that shows files and directories and lets the user navigate the file system is hard to write, and you definitely don't want to reinvent that wheel. Fortunately, Swing provides a JFileChooser class that allows you to display a file dialog box similar to the one that most native applications use. JFileChooser dialogs are always modal. Note that the JFileChooser class is not a subclass of JDialog. Instead of calling setVisible(true), call showOpenDialog to display a dialog for opening a file, or call showSaveDialog to display a dialog for saving a file. The button for accepting a file is then automatically labeled Open or Save. You can also supply your own button label with the showDialog method. Figure 11.33 shows an example of the file chooser dialog box.

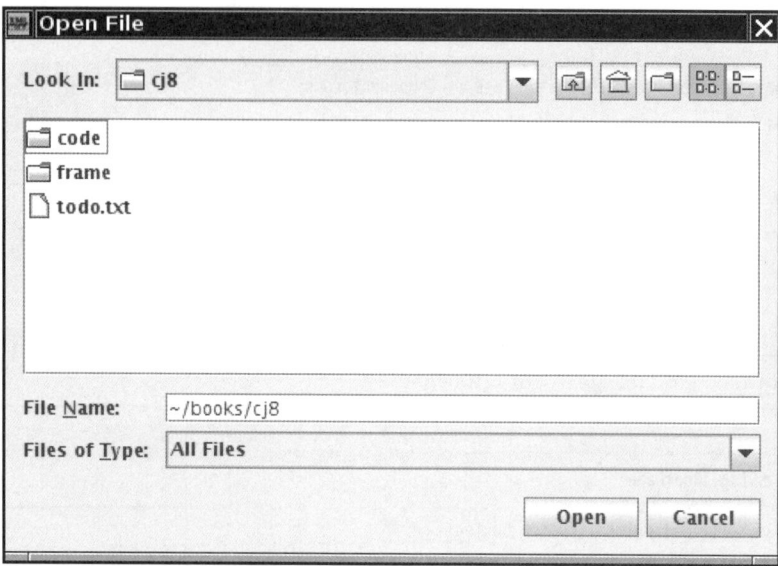

Figure 11.33 A file chooser dialog box

Here are the steps to put up a file dialog box and recover what the user chooses from the box:

1. Make a JFileChooser object. Unlike the constructor for the JDialog class, you do not supply the parent component. This allows you to reuse a file chooser dialog with multiple frames.

    For example:

    ```
    var chooser = new JFileChooser();
    ```

 TIP: Reusing a file chooser object is a good idea because the JFileChooser constructor can be quite slow, especially on Windows when the user has many mapped network drives.

---

2. Set the directory by calling the setCurrentDirectory method.

    For example, to use the current working directory

    ```
    chooser.setCurrentDirectory(new File("."));
    ```

    you need to supply a File object. File objects are explained in detail in Chapter 2 of Volume II. All you need to know for now is that the constructor File(String filename) turns a file or directory name into a File object.

3. If you have a default file name that you expect the user to choose, supply it with the `setSelectedFile` method:

    ```
    chooser.setSelectedFile(new File(filename));
    ```

4. To enable the user to select multiple files in the dialog, call the `setMultiSelectionEnabled` method. This is, of course, entirely optional and not all that common.

    ```
    chooser.setMultiSelectionEnabled(true);
    ```

5. If you want to restrict the display of files in the dialog to those of a particular type (for example, all files with extension .gif), you need to set a *file filter*. We discuss file filters later in this section.

6. By default, a user can select only files with a file chooser. If you want the user to select directories, use the `setFileSelectionMode` method. Call it with `JFileChooser.FILES_ONLY` (the default), `JFileChooser.DIRECTORIES_ONLY`, or `JFileChooser.FILES_AND_DIRECTORIES`.

7. Show the dialog box by calling the `showOpenDialog` or `showSaveDialog` method. You must supply the parent component in these calls:

    ```
    int result = chooser.showOpenDialog(parent);
    ```

    or

    ```
    int result = chooser.showSaveDialog(parent);
    ```

    The only difference between these calls is the label of the "approve button"—the button that the user clicks to finish the file selection. You can also call the `showDialog` method and pass an explicit text for the approve button:

    ```
    int result = chooser.showDialog(parent, "Select");
    ```

    These calls return only when the user has approved, canceled, or dismissed the file dialog. The return value is `JFileChooser.APPROVE_OPTION`, `JFileChooser.CANCEL_OPTION`, or `JFileChooser.ERROR_OPTION`.

8. Get the selected file or files with the `getSelectedFile()` or `getSelectedFiles()` method. These methods return either a single `File` object or an array of `File` objects. If you just need the name of the file object, call its `getPath` method. For example:

    ```
    String filename = chooser.getSelectedFile().getPath();
    ```

For the most part, these steps are simple. The major difficulty with using a file dialog is to specify a subset of files from which the user should choose. For example, suppose the user should choose a GIF image file. Then, the file chooser should only display files with the extension .gif. It should also give the user some kind of feedback that the displayed files are of a particular category, such as "GIF Images." But the situation can be more complex. If the user should choose a JPEG image file, the extension can be either .jpg or .jpeg. Instead of a way to codify these complexities, the designers of the file chooser provided a more elegant mechanism: to restrict the displayed files, supply an object that extends the abstract class javax.swing.filechooser.FileFilter. The file chooser passes each file to the file filter and displays only those files that the filter accepts.

At the time of this writing, two such subclasses are supplied: the default filter that accepts all files, and a filter that accepts all files with a given extension. However, it is easy to write ad-hoc file filters. Simply implement the two abstract methods of the FileFilter superclass:

```
public boolean accept(File f);
public String getDescription();
```

The first method tests whether a file should be accepted. The second method returns a description of the file type that can be displayed in the file chooser dialog.

 NOTE: An unrelated FileFilter interface in the java.io package has a single method, boolean accept(File f). It is used in the listFiles method of the File class to list files in a directory. We do not know why the designers of Swing didn't extend this interface—perhaps the Java class library has now become so complex that even the programmers at Sun were no longer aware of all the standard classes and interfaces.

You will need to resolve the name conflict between these two identically named types if you import both the java.io and the javax.swing.filechooser packages. The simplest remedy is to import javax.swing.filechooser.FileFilter, not javax.swing.filechooser.*.

Once you have a file filter object, use the setFileFilter method of the JFileChooser class to install it into the file chooser object:

```
chooser.setFileFilter(new FileNameExtensionFilter("Image files", "gif", "jpg"));
```

You can install multiple filters to the file chooser by calling

```
chooser.addChoosableFileFilter(filter1);
chooser.addChoosableFileFilter(filter2);
. . .
```

The user selects a filter from the combo box at the bottom of the file dialog. By default, the "All files" filter is always present in the combo box. This is a good idea—just in case a user of your program needs to select a file with a nonstandard extension. However, if you want to suppress the "All files" filter, call

```
chooser.setAcceptAllFileFilterUsed(false)
```

 CAUTION: If you reuse a single file chooser for loading and saving different file types, call

```
chooser.resetChoosableFilters()
```

to clear any old file filters before adding new ones.

Finally, you can customize the file chooser by providing special icons and file descriptions for each file that the file chooser displays. Do this by supplying an object of a class extending the FileView class in the javax.swing.filechooser package. This is definitely an advanced technique. Normally, you don't need to supply a file view—the pluggable look-and-feel supplies one for you. But if you want to show different icons for special file types, you can install your own file view. You need to extend the FileView class and implement five methods:

```
Icon getIcon(File f)
String getName(File f)
String getDescription(File f)
String getTypeDescription(File f)
Boolean isTraversable(File f)
```

Then, use the setFileView method to install your file view into the file chooser.

The file chooser calls your methods for each file or directory that it wants to display. If your method returns null for the icon, name, or description, the file chooser then consults the default file view of the look-and-feel. That is good, because it means you need to deal only with the file types for which you want to do something different.

The file chooser calls the isTraversable method to decide whether to open a directory when a user clicks on it. Note that this method returns a Boolean object, not a boolean value! This seems weird, but it is actually convenient—if you aren't interested in deviating from the default file view, just return null. The file chooser will then consult the default file view. In other words, the method returns a Boolean to let you choose among three options: true (Boolean.TRUE), false (Boolean.FALSE), or don't care (null).

The example program contains a simple file view class. That class shows a particular icon whenever a file matches a file filter. We use it to display a palette icon for all image files.

```
class FileIconView extends FileView
{
   private FileFilter filter;
   private Icon icon;

   public FileIconView(FileFilter aFilter, Icon anIcon)
   {
      filter = aFilter;
      icon = anIcon;
   }

   public Icon getIcon(File f)
   {
      if (!f.isDirectory() && filter.accept(f))
         return icon;
      else return null;
   }
}
```

Install this file view into your file chooser with the setFileView method:

```
chooser.setFileView(new FileIconView(filter,
   new ImageIcon("palette.gif")));
```

The file chooser will then show the palette icon next to all files that pass the filter and use the default file view to show all other files. Naturally, we use the same filter that we set in the file chooser.

Finally, you can customize a file dialog by adding an *accessory* component. For example, Figure 11.34 shows a preview accessory next to the file list. This accessory displays a thumbnail view of the currently selected file.

An accessory can be any Swing component. In our case, we extend the JLabel class and set its icon to a scaled copy of the graphics image:

```
class ImagePreviewer extends JLabel
{
   public ImagePreviewer(JFileChooser chooser)
   {
      setPreferredSize(new Dimension(100, 100));
      setBorder(BorderFactory.createEtchedBorder());
   }

   public void loadImage(File f)
   {
```

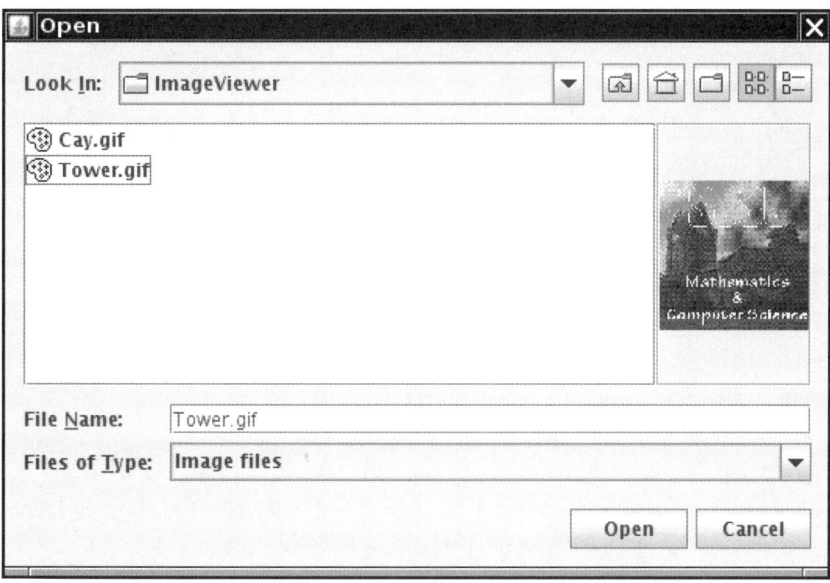

Figure 11.34 A file dialog with a preview accessory

```
      var icon = new ImageIcon(f.getPath());
      if(icon.getIconWidth() > getWidth())
         icon = new ImageIcon(icon.getImage().getScaledInstance(
            getWidth(), -1, Image.SCALE_DEFAULT));
      setIcon(icon);
      repaint();
   }
}
```

There is just one challenge. We want to update the preview image whenever the user selects a different file. The file chooser uses the "JavaBeans" mechanism of notifying interested listeners whenever one of its properties changes. The selected file is a property that you can monitor by installing a PropertyChangeListener. Here is the code that you need to trap the notifications:

```
chooser.addPropertyChangeListener(event -> {
   if (event.getPropertyName() == JFileChooser.SELECTED_FILE_CHANGED_PROPERTY)
   {
      var newFile = (File) event.getNewValue();
      // update the accessory
      . . .
   }
});
```

---

javax.swing.JFileChooser 1.2

- JFileChooser()

  creates a file chooser dialog box that can be used for multiple frames.

- void setCurrentDirectory(File dir)

  sets the initial directory for the file dialog box.

- void setSelectedFile(File file)
- void setSelectedFiles(File[] file)

  sets the default file choice for the file dialog box.

- void setMultiSelectionEnabled(boolean b)

  sets or clears the multiple selection mode.

- void setFileSelectionMode(int mode)

  lets the user select files only (the default), directories only, or both files and directories. The mode parameter is one of JFileChooser.FILES_ONLY, JFileChooser.DIRECTORIES_ONLY, and JFileChooser.FILES_AND_DIRECTORIES.

- int showOpenDialog(Component parent)
- int showSaveDialog(Component parent)
- int showDialog(Component parent, String approveButtonText)

  shows a dialog in which the approve button is labeled "Open", "Save", or with the approveButtonText string. Returns APPROVE_OPTION, CANCEL_OPTION (if the user selected the cancel button or dismissed the dialog), or ERROR_OPTION (if an error occurred).

- File getSelectedFile()
- File[] getSelectedFiles()

  gets the file or files that the user selected (or returns null if the user didn't select any file).

- void setFileFilter(FileFilter filter)

  sets the file mask for the file dialog box. All files for which filter.accept returns true will be displayed. Also, adds the filter to the list of choosable filters.

- void addChoosableFileFilter(FileFilter filter)

  adds a file filter to the list of choosable filters.

- void setAcceptAllFileFilterUsed(boolean b)

  includes or suppresses an "All files" filter in the filter combo box.

(Continues)

---

javax.swing.JFileChooser 1.2 *(Continued)*

- void resetChoosableFileFilters()

  clears the list of choosable filters. Only the "All files" filter remains unless it is explicitly suppressed.

- void setFileView(FileView view)

  sets a file view to provide information about the files that the file chooser displays.

- void setAccessory(JComponent component)

  sets an accessory component.

---

javax.swing.filechooser.FileFilter 1.2

- boolean accept(File f)

  returns true if the file chooser should display this file.

- String getDescription()

  returns a description of this file filter—for example, "Image files (*.gif,*.jpeg)".

---

javax.swing.filechooser.FileNameExtensionFilter 6

- FileNameExtensionFilter(String description, String... extensions)

  constructs a file filter with the given description that accepts all directories and all files whose names end in a period followed by one of the given extension strings.

---

javax.swing.filechooser.FileView 1.2

- String getName(File f)

  returns the name of the file f, or null. Normally, this method simply returns f.getName().

- String getDescription(File f)

  returns a human-readable description of the file f, or null. For example, if f is an HTML document, this method might return its title.

(Continues)

`javax.swing.filechooser.FileView` 1.2 *(Continued)*

- `String getTypeDescription(File f)`

  returns a human-readable description of the type of the file f, or null. For example, if f is an HTML document, this method might return a string "Hypertext document".

- `Icon getIcon(File f)`

  returns an icon for the file f, or null. For example, if f is a JPEG file, this method might return a thumbnail icon.

- `Boolean isTraversable(File f)`

  returns Boolean.TRUE if f is a directory that the user can open. This method might return Boolean.FALSE if a directory is conceptually a compound document. Like all FileView methods, this method can return null to signify that the file chooser should consult the default view instead.

This ends our discussion of Swing programming. Turn to Volume II for more advanced Swing components and sophisticated graphics techniques.

CHAPTER

12

Concurrency

In this chapter

You are probably familiar with *multitasking*—your operating system's ability to have more than one program working at what seems like the same time. For example, you can print while editing or downloading your email. Nowadays, you are likely to have a computer with more than one CPU, but the number of concurrently executing processes is not limited by the number of CPUs. The operating system assigns CPU time slices to each process, giving the impression of parallel activity.

Multithreaded programs extend the idea of multitasking by taking it one level lower: Individual programs will appear to do multiple tasks at the same time. Each task is executed in a *thread*, which is short for thread of control. Programs that can run more than one thread at once are said to be *multithreaded*.

So, what is the difference between multiple *processes* and multiple *threads*? The essential difference is that while each process has a complete set of its own variables, threads share the same data. This sounds somewhat risky, and indeed it can be, as you will see later in this chapter. However, shared variables make communication between threads more efficient and easier to program than interprocess communication. Moreover, on some operating systems, threads are more "lightweight" than processes—it takes less overhead to create and destroy individual threads than it does to launch new processes.

Multithreading is extremely useful in practice. For example, a browser should be able to simultaneously download multiple images. A web server needs to be able to serve concurrent requests. Graphical user interface (GUI) programs have a separate thread for gathering user interface events from the host operating environment. This chapter shows you how to add multithreading capability to your Java applications.

Fair warning: Concurrent programming can get very complex. In this chapter, we cover all the tools that an application programmer is likely to need. However, for more intricate system-level programming, we suggest that you turn to a more advanced reference, such as *Java Concurrency in Practice* by Brian Goetz et al. (Addison-Wesley Professional, 2006).

12.1 What Are Threads?

Let us start by looking at a simple program that uses two threads. This program moves money between bank accounts. We make use of a Bank class that stores the balances of a given number of accounts. The transfer method transfers an amount from one account to another. See Listing 12.2 for the implementation.

In the first thread, we will move money from account 0 to account 1. The second thread moves money from account 2 to account 3.

Here is a simple procedure for running a task in a separate thread:

1. Place the code for the task into the run method of a class that implements the Runnable interface. That interface is very simple, with a single method:

   ```
   public interface Runnable
   {
      void run();
   }
   ```

Since Runnable is a functional interface, you can make an instance with a lambda expression:

```
Runnable r = () -> { task code };
```

2. Construct a Thread object from the Runnable:

```
var t = new Thread(r);
```

3. Start the thread:

```
t.start();
```

To make a separate thread for transferring money, we only need to place the code for the transfer inside the run method of a Runnable, and then start a thread:

```
Runnable r = () -> {
   try
   {
      for (int i = 0; i < STEPS; i++)
      {
         double amount = MAX_AMOUNT * Math.random();
         bank.transfer(0, 1, amount);
         Thread.sleep((int) (DELAY * Math.random()));
      }
   }
   catch (InterruptedException e)
   {
   }
};
var t = new Thread(r);
t.start();
```

For a given number of steps, this thread transfers a random amount, and then sleeps for a random delay.

We need to catch an InterruptedException that the sleep method threatens to throw. We will discuss this exception in Section 12.3.1, "Interrupting Threads," on p. 743. Typically, interruption is used to request that a thread terminates. Accordingly, our run method exits when an InterruptedException occurs.

Our program starts a second thread as well that moves money from account 2 to account 3. When you run this program, you get a printout like this:

```
Thread[Thread-1,5,main]    606.77 from 2 to 3 Total Balance:  400000.00
Thread[Thread-0,5,main]     98.99 from 0 to 1 Total Balance:  400000.00
Thread[Thread-1,5,main]    476.78 from 2 to 3 Total Balance:  400000.00
Thread[Thread-0,5,main]    653.64 from 0 to 1 Total Balance:  400000.00
Thread[Thread-1,5,main]    807.14 from 2 to 3 Total Balance:  400000.00
Thread[Thread-0,5,main]    481.49 from 0 to 1 Total Balance:  400000.00
Thread[Thread-0,5,main]    203.73 from 0 to 1 Total Balance:  400000.00
```

```
Thread[Thread-1,5,main]    111.76 from 2 to 3 Total Balance:  400000.00
Thread[Thread-1,5,main]    794.88 from 2 to 3 Total Balance:  400000.00
. . .
```

As you can see, the output of the two threads is interleaved, showing that they run concurrently. In fact, sometimes the output is a little messier when two output lines are interleaved.

That's all there is to it! You now know how to run tasks concurrently. The remainder of this chapter tells you how to control the interaction between threads.

The complete code is shown in Listing 12.1.

 NOTE: You can also define a thread by forming a subclass of the Thread class, like this:

```
class MyThread extends Thread
{
   public void run()
   {
      task code
   }
}
```

Then you construct an object of the subclass and call its start method. However, this approach is no longer recommended. You should decouple the *task* that is to be run in parallel from the *mechanism* of running it. If you have many tasks, it is too expensive to create a separate thread for each of them. Instead, you can use a thread pool—see Section 12.6.2, "Executors," on p. 802.

 CAUTION: Do *not* call the run method of the Thread class or the Runnable object. Calling the run method directly merely executes the task in the *same* thread—no new thread is started. Instead, call the Thread.start method. It creates a new thread that executes the run method.

Listing 12.1 threads/ThreadTest.java

```
1 package threads;
2
3 /**
4  * @version 1.30 2004-08-01
5  * @author Cay Horstmann
6  */
```

```
 7  public class ThreadTest
 8  {
 9     public static final int DELAY = 10;
10     public static final int STEPS = 100;
11     public static final double MAX_AMOUNT = 1000;
12
13     public static void main(String[] args)
14     {
15        var bank = new Bank(4, 100000);
16        Runnable task1 = () ->
17        {
18           try
19           {
20              for (int i = 0; i < STEPS; i++)
21              {
22                 double amount = MAX_AMOUNT * Math.random();
23                 bank.transfer(0, 1, amount);
24                 Thread.sleep((int) (DELAY * Math.random()));
25              }
26           }
27           catch (InterruptedException e)
28           {
29           }
30        };
31
32        Runnable task2 = () ->
33        {
34           try
35           {
36              for (int i = 0; i < STEPS; i++)
37              {
38                 double amount = MAX_AMOUNT * Math.random();
39                 bank.transfer(2, 3, amount);
40                 Thread.sleep((int) (DELAY * Math.random()));
41              }
42           }
43           catch (InterruptedException e)
44           {
45           }
46        };
47
48        new Thread(task1).start();
49        new Thread(task2).start();
50     }
51  }
```

Listing 12.2 threads/Bank.java

```java
1  package threads;
2
3  import java.util.*;
4
5  /**
6   * A bank with a number of bank accounts.
7   */
8  public class Bank
9  {
10    private final double[] accounts;
11
12    /**
13     * Constructs the bank.
14     * @param n the number of accounts
15     * @param initialBalance the initial balance for each account
16     */
17    public Bank(int n, double initialBalance)
18    {
19       accounts = new double[n];
20       Arrays.fill(accounts, initialBalance);
21    }
22
23    /**
24     * Transfers money from one account to another.
25     * @param from the account to transfer from
26     * @param to the account to transfer to
27     * @param amount the amount to transfer
28     */
29    public void transfer(int from, int to, double amount)
30    {
31       if (accounts[from] < amount) return;
32       System.out.print(Thread.currentThread());
33       accounts[from] -= amount;
34       System.out.printf(" %10.2f from %d to %d", amount, from, to);
35       accounts[to] += amount;
36       System.out.printf(" Total Balance: %10.2f%n", getTotalBalance());
37    }
38
39    /**
40     * Gets the sum of all account balances.
41     * @return the total balance
42     */
43    public double getTotalBalance()
44    {
45       double sum = 0;
46
47       for (double a : accounts)
48          sum += a;
```

```
49
50     return sum;
51   }
52
53   /**
54    * Gets the number of accounts in the bank.
55    * @return the number of accounts
56    */
57   public int size()
58   {
59      return accounts.length;
60   }
61 }
```

---

java.lang.Thread 1.0

- Thread(Runnable target)

  constructs a new thread that calls the run() method of the specified target.

- void start()

  starts this thread, causing the run() method to be called. This method will return immediately. The new thread runs concurrently.

- void run()

  calls the run method of the associated Runnable.

- static void sleep(long millis)

  sleeps for the given number of milliseconds.

---

java.lang.Runnable 1.0

- void run()

  must be overridden and supplied with instructions for the task that you want to have executed.

12.2 Thread States

Threads can be in one of six states:

- New
- Runnable
- Blocked
- Waiting

- Timed waiting
- Terminated

Each of these states is explained in the sections that follow.

To determine the current state of a thread, simply call the getState method.

12.2.1 New Threads

When you create a thread with the new operator—for example, new Thread(r)—the thread is not yet running. This means that it is in the *new* state. When a thread is in the new state, the program has not started executing code inside of it. A certain amount of bookkeeping needs to be done before a thread can run.

12.2.2 Runnable Threads

Once you invoke the start method, the thread is in the *runnable* state. A runnable thread may or may not actually be running. It is up to the operating system to give the thread time to run. (The Java specification does not call this a separate state, though. A running thread is still in the runnable state.)

Once a thread is running, it doesn't necessarily keep running. In fact, it is desirable that running threads occasionally pause so that other threads have a chance to run. The details of thread scheduling depend on the services that the operating system provides. Preemptive scheduling systems give each runnable thread a slice of time to perform its task. When that slice of time is exhausted, the operating system *preempts* the thread and gives another thread an opportunity to work (see Figure 12.2). When selecting the next thread, the operating system takes into account the thread *priorities*—see Section 12.3.5, "Thread Priorities," on p. 749 for more information.

All modern desktop and server operating systems use preemptive scheduling. However, small devices such as cell phones may use cooperative scheduling. In such a device, a thread loses control only when it calls the yield method, or when it is blocked or waiting.

On a machine with multiple processors, each processor can run a thread, and you can have multiple threads run in parallel. Of course, if there are more threads than processors, the scheduler still has to do time slicing.

Always keep in mind that a runnable thread may or may not be running at any given time. (This is why the state is called "runnable" and not "running.")

---

java.lang.Thread 1.0

- static void yield()

  causes the currently executing thread to yield to another thread. Note that this is a static method.

12.2.3 Blocked and Waiting Threads

When a thread is blocked or waiting, it is temporarily inactive. It doesn't execute any code and consumes minimal resources. It is up to the thread scheduler to reactivate it. The details depend on how the inactive state was reached.

- When the thread tries to acquire an intrinsic object lock (but not a Lock in the java.util.concurrent library) that is currently held by another thread, it becomes *blocked*. (We discuss java.util.concurrent locks in Section 12.4.3, "Lock Objects," on p. 755 and intrinsic object locks in Section 12.4.5, "The synchronized Keyword," on p. 764.) The thread becomes unblocked when all other threads have relinquished the lock and the thread scheduler has allowed this thread to hold it.

- When the thread waits for another thread to notify the scheduler of a condition, it enters the *waiting* state. We discuss conditions in Section 12.4.4, "Condition Objects," on p. 758. This happens by calling the Object.wait or Thread.join method, or by waiting for a Lock or Condition in the java.util.concurrent library. In practice, the difference between the blocked and waiting state is not significant.

- Several methods have a timeout parameter. Calling them causes the thread to enter the *timed waiting* state. This state persists either until the timeout expires or the appropriate notification has been received. Methods with timeout include Thread.sleep and the timed versions of Object.wait, Thread.join, Lock.tryLock, and Condition.await.

Figure 12.1 shows the states that a thread can have and the possible transitions from one state to another. When a thread is blocked or waiting (or, of course, when it terminates), another thread will be scheduled to run. When a thread is reactivated (for example, because its timeout has expired or it has succeeded in acquiring a lock), the scheduler checks to see if it has a higher priority than the currently running threads. If so, it preempts one of the current threads and picks a new thread to run.

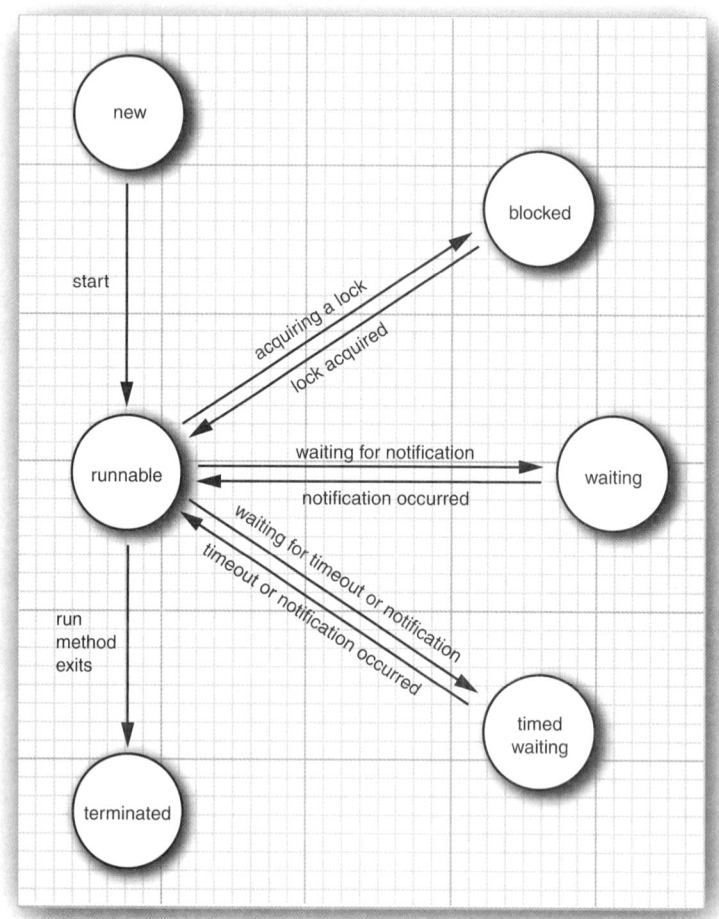

Figure 12.1 Thread states

12.2.4 Terminated Threads

A thread is terminated for one of two reasons:

- It dies a natural death because the run method exits normally.
- It dies abruptly because an uncaught exception terminates the run method.

In particular, you can kill a thread by invoking its stop method. That method throws a ThreadDeath error object that kills the thread. However, the stop method is deprecated, and you should never call it in your own code.

---

java.lang.Thread 1.0

- void join()

  waits for the specified thread to terminate.

- void join(long millis)

  waits for the specified thread to die or for the specified number of milliseconds to pass.

- Thread.State getState() 5

  gets the state of this thread: one of NEW, RUNNABLE, BLOCKED, WAITING, TIMED_WAITING, or TERMINATED.

- void stop()

  stops the thread. This method is deprecated.

- void suspend()

  suspends this thread's execution. This method is deprecated.

- void resume()

  resumes this thread. This method is only valid after suspend() has been invoked. This method is deprecated.

12.3 Thread Properties

In the following sections, we discuss miscellaneous properties of threads: the interrupted status, daemon threads, handlers for uncaught exceptions, as well as some legacy features that you should not use.

12.3.1 Interrupting Threads

A thread terminates when its run method returns—by executing a return statement, after executing the last statement in the method body, or if an exception occurs that is not caught in the method. In the initial release of Java, there also was a stop method that another thread could call to terminate a thread. However, that method is now deprecated. We discuss the reason in Section 12.4.13, "Why the stop and suspend Methods Are Deprecated," on p. 779.

Other than with the deprecated stop method, there is no way to *force* a thread to terminate. However, the interrupt method can be used to *request* termination of a thread.

When the interrupt method is called on a thread, the *interrupted status* of the thread is set. This is a boolean flag that is present in every thread. Each thread should occasionally check whether it has been interrupted.

To find out whether the interrupted status was set, first call the static Thread.currentThread method to get the current thread, and then call the isInterrupted method:

```
while (!Thread.currentThread().isInterrupted() && more work to do)
{
    do more work
}
```

However, if a thread is blocked, it cannot check the interrupted status. This is where the InterruptedException comes in. When the interrupt method is called on a thread that blocks on a call such as sleep or wait, the blocking call is terminated by an InterruptedException. (There are blocking I/O calls that cannot be interrupted; you should consider interruptible alternatives. See Chapters 2 and 4 of Volume II for details.)

There is no language requirement that a thread which is interrupted should terminate. Interrupting a thread simply grabs its attention. The interrupted thread can decide how to react to the interruption. Some threads are so important that they should handle the exception and continue. But quite commonly, a thread will simply want to interpret an interruption as a request for termination. The run method of such a thread has the following form:

```
Runnable r = () -> {
    try
    {
        . . .
        while (!Thread.currentThread().isInterrupted() && more work to do)
        {
            do more work
        }
    }
    catch(InterruptedException e)
    {
        // thread was interrupted during sleep or wait
    }
    finally
    {
        cleanup, if required
    }
    // exiting the run method terminates the thread
};
```

The isInterrupted check is neither necessary nor useful if you call the sleep method (or another interruptible method) after every work iteration. If you call the sleep method when the interrupted status is set, it doesn't sleep. Instead, it clears the status (!) and throws an InterruptedException. Therefore, if

your loop calls sleep, don't check the interrupted status. Instead, catch the
InterruptedException, like this:

```
Runnable r = () -> {
   try
   {
      . . .
      while (more work to do)
      {
         do more work
         Thread.sleep(delay);
      }
   }
   catch(InterruptedException e)
   {
      // thread was interrupted during sleep
   }
   finally
   {
      cleanup, if required
   }
   // exiting the run method terminates the thread
};
```

 NOTE: There are two very similar methods, interrupted and isInterrupted. The
interrupted method is a static method that checks whether the *current* thread
has been interrupted. Furthermore, calling the interrupted method *clears* the interrupted status of the thread. On the other hand, the isInterrupted method is
an instance method that you can use to check whether any thread has been
interrupted. Calling it does not change the interrupted status.

You'll find lots of published code in which the InterruptedException is squelched
at a low level, like this:

```
void mySubTask()
{
   . . .
   try { sleep(delay); }
   catch (InterruptedException e) {} // don't ignore!
   . . .
}
```

Don't do that! If you can't think of anything good to do in the catch clause,
you still have two reasonable choices:

• In the catch clause, call Thread.currentThread().interrupt() to set the interrupted
  status. Then the caller can test it.

```
void mySubTask()
{
    . . .
    try { sleep(delay); }
    catch (InterruptedException e) { Thread.currentThread().interrupt(); }
    . . .
}
```

- Or, even better, tag your method with throws InterruptedException and drop the try block. Then the caller (or, ultimately, the run method) can catch it.

```
void mySubTask() throws InterruptedException
{
    . . .
    sleep(delay);
    . . .
}
```

---

java.lang.Thread 1.0

- void interrupt()

  sends an interrupt request to a thread. The interrupted status of the thread is set to true. If the thread is currently blocked by a call to sleep, then an InterruptedException is thrown.

- static boolean interrupted()

  tests whether the *current* thread (that is, the thread that is executing this instruction) has been interrupted. Note that this is a static method. The call has a side effect—it resets the interrupted status of the current thread to false.

- boolean isInterrupted()

  tests whether a thread has been interrupted. Unlike the static interrupted method, this call does not change the interrupted status of the thread.

- static Thread currentThread()

  returns the Thread object representing the currently executing thread.

12.3.2 Daemon Threads

You can turn a thread into a *daemon thread* by calling

```
t.setDaemon(true);
```

There is nothing demonic about such a thread. A daemon is simply a thread that has no other role in life than to serve others. Examples are timer threads that send regular "timer ticks" to other threads or threads that clean up stale cache entries. When only daemon threads remain, the virtual machine exits. There is no point in keeping the program running if all remaining threads are daemons.

---

java.lang.Thread 1.0

- void setDaemon(boolean isDaemon)

  marks this thread as a daemon thread or a user thread. This method must be called before the thread is started.

12.3.3 Thread Names

By default, threads have catchy names such as Thread-2. You can set any name with the setName method:

```
var t = new Thread(runnable);
t.setName("Web crawler");
```

That can be useful in thread dumps.

12.3.4 Handlers for Uncaught Exceptions

The run method of a thread cannot throw any checked exceptions, but it can be terminated by an unchecked exception. In that case, the thread dies.

However, there is no catch clause to which the exception can be propagated. Instead, just before the thread dies, the exception is passed to a handler for uncaught exceptions.

The handler must belong to a class that implements the Thread.UncaughtExceptionHandler interface. That interface has a single method,

```
void uncaughtException(Thread t, Throwable e)
```

You can install a handler into any thread with the setUncaughtExceptionHandler method. You can also install a default handler for all threads with the static method setDefaultUncaughtExceptionHandler of the Thread class. A replacement handler might use the logging API to send reports of uncaught exceptions into a log file.

If you don't install a default handler, the default handler is null. However, if you don't install a handler for an individual thread, the handler is the thread's ThreadGroup object.

NOTE: A thread group is a collection of threads that can be managed together. By default, all threads that you create belong to the same thread group, but it is possible to establish other groupings. Since there are now better features for operating on collections of threads, we recommend that you do not use thread groups in your programs.

The `ThreadGroup` class implements the `Thread.UncaughtExceptionHandler` interface. Its `uncaughtException` method takes the following action:

1. If the thread group has a parent, then the `uncaughtException` method of the parent group is called.

2. Otherwise, if the `Thread.getDefaultUncaughtExceptionHandler` method returns a non-`null` handler, it is called.

3. Otherwise, if the `Throwable` is an instance of `ThreadDeath`, nothing happens.

4. Otherwise, the name of the thread and the stack trace of the `Throwable` are printed on `System.err`.

That is the stack trace that you have undoubtedly seen many times in your programs.

---

java.lang.Thread 1.0

- `static void setDefaultUncaughtExceptionHandler(Thread.UncaughtExceptionHandler handler)` 5
- `static Thread.UncaughtExceptionHandler getDefaultUncaughtExceptionHandler()` 5

  sets or gets the default handler for uncaught exceptions.

- `void setUncaughtExceptionHandler(Thread.UncaughtExceptionHandler handler)` 5
- `Thread.UncaughtExceptionHandler getUncaughtExceptionHandler()` 5

  sets or gets the handler for uncaught exceptions. If no handler is installed, the thread group object is the handler.

---

java.lang.Thread.UncaughtExceptionHandler 5

- `void uncaughtException(Thread t, Throwable e)`

  defined to log a custom report when a thread is terminated with an uncaught exception.

---

java.lang.ThreadGroup 1.0

- void uncaughtException(Thread t, Throwable e)

  calls this method of the parent thread group if there is a parent, or calls the default handler of the Thread class if there is a default handler, or otherwise prints a stack trace to the standard error stream. (However, if e is a ThreadDeath object, the stack trace is suppressed. ThreadDeath objects are generated by the deprecated stop method.)

12.3.5 Thread Priorities

In the Java programming language, every thread has a *priority*. By default, a thread inherits the priority of the thread that constructed it. You can increase or decrease the priority of any thread with the setPriority method. You can set the priority to any value between MIN_PRIORITY (defined as 1 in the Thread class) and MAX_PRIORITY (defined as 10). NORM_PRIORITY is defined as 5.

Whenever the thread scheduler has a chance to pick a new thread, it prefers threads with higher priority. However, thread priorities are *highly system-dependent*. When the virtual machine relies on the thread implementation of the host platform, the Java thread priorities are mapped to the priority levels of the host platform, which may have more or fewer thread priority levels.

For example, Windows has seven priority levels. Some of the Java priorities will map to the same operating system level. In the Oracle JVM for Linux, thread priorities are ignored altogether—all threads have the same priority.

Thread priorities may have been useful in early versions of Java that didn't use operating systems threads. You should not use them nowadays.

---

java.lang.Thread 1.0

- void setPriority(int newPriority)

  sets the priority of this thread. The priority must be between Thread.MIN_PRIORITY and Thread.MAX_PRIORITY. Use Thread.NORM_PRIORITY for normal priority.

- static int MIN_PRIORITY

  is the minimum priority that a Thread can have. The minimum priority value is 1.

- static int NORM_PRIORITY

  is the default priority of a Thread. The default priority is 5.

- static int MAX_PRIORITY

  is the maximum priority that a Thread can have. The maximum priority value is 10.

12.4 Synchronization

In most practical multithreaded applications, two or more threads need to share access to the same data. What happens if two threads have access to the same object and each calls a method that modifies the state of the object? As you might imagine, the threads can step on each other's toes. Depending on the order in which the data were accessed, corrupted objects can result. Such a situation is often called a *race condition.*

12.4.1 An Example of a Race Condition

To avoid corruption of shared data by multiple threads, you must learn how to *synchronize the access.* In this section, you'll see what happens if you do not use synchronization. In the next section, you'll see how to synchronize data access.

In the next test program, we continue working with our simulated bank. Unlike the example in Section 12.1, "What Are Threads?," on p. 734, we randomly select the source and destination of the transfer. Since this will cause problems, let us look more carefully at the code for the `transfer` method of the Bank class.

```
public void transfer(int from, int to, double amount)
   // CAUTION: unsafe when called from multiple threads
{
   System.out.print(Thread.currentThread());
   accounts[from] -= amount;
   System.out.printf(" %10.2f from %d to %d", amount, from, to);
   accounts[to] += amount;
   System.out.printf(" Total Balance: %10.2f%n", getTotalBalance());
}
```

Here is the code for the `Runnable` instances. The `run` method keeps moving money out of a given bank account. In each iteration, the `run` method picks a random target account and a random amount, calls `transfer` on the bank object, and then sleeps.

```
Runnable r = () -> {
   try
   {
      while (true)
      {
         int toAccount = (int) (bank.size() * Math.random());
         double amount = MAX_AMOUNT * Math.random();
         bank.transfer(fromAccount, toAccount, amount);
         Thread.sleep((int) (DELAY * Math.random()));
      }
   }
```

```
catch (InterruptedException e)
{
}
};
```

When this simulation runs, we do not know how much money is in any one bank account at any time. But we do know that the total amount of money in all the accounts should remain unchanged because all we do is move money from one account to another.

At the end of each transaction, the transfer method recomputes the total and prints it.

This program never finishes. Just press Ctrl+C to kill the program.

Here is a typical printout:

```
. . .
Thread[Thread-11,5,main]    588.48 from 11 to 44 Total Balance:  100000.00
Thread[Thread-12,5,main]    976.11 from 12 to 22 Total Balance:  100000.00
Thread[Thread-14,5,main]    521.51 from 14 to 22 Total Balance:  100000.00
Thread[Thread-13,5,main]    359.89 from 13 to 81 Total Balance:  100000.00

. . .
Thread[Thread-36,5,main]    401.71 from 36 to 73 Total Balance:   99291.06
Thread[Thread-35,5,main]    691.46 from 35 to 77 Total Balance:   99291.06
Thread[Thread-37,5,main]     78.64 from 37 to 3 Total Balance:   99291.06
Thread[Thread-34,5,main]    197.11 from 34 to 69 Total Balance:   99291.06
Thread[Thread-36,5,main]     85.96 from 36 to 4 Total Balance:   99291.06
. . .
Thread[Thread-4,5,main]Thread[Thread-33,5,main]      7.31 from 31 to 32 Total Balance:
99979.24
        627.50 from 4 to 5 Total Balance:  99979.24                    ·
. . .
```

As you can see, something is very wrong. For a few transactions, the bank balance remains at $100,000, which is the correct total for 100 accounts of $1,000 each. But after some time, the balance changes slightly. When you run this program, errors may happen quickly, or it may take a very long time for the balance to become corrupted. This situation does not inspire confidence, and you would probably not want to deposit your hard-earned money in such a bank.

See if you can spot the problems with the code in Listing 12.3 and the Bank class in Listing 12.2. We will unravel the mystery in the next section.

Listing 12.3 unsynch/UnsynchBankTest.java

```java
1  package unsynch;
2
3  /**
4   * This program shows data corruption when multiple threads access a data structure.
5   * @version 1.32 2018-04-10
6   * @author Cay Horstmann
7   */
8  public class UnsynchBankTest
9  {
10     public static final int NACCOUNTS = 100;
11     public static final double INITIAL_BALANCE = 1000;
12     public static final double MAX_AMOUNT = 1000;
13     public static final int DELAY = 10;
14
15     public static void main(String[] args)
16     {
17        var bank = new Bank(NACCOUNTS, INITIAL_BALANCE);
18        for (int i = 0; i < NACCOUNTS; i++)
19        {
20           int fromAccount = i;
21           Runnable r = () -> {
22              try
23              {
24                 while (true)
25                 {
26                    int toAccount = (int) (bank.size() * Math.random());
27                    double amount = MAX_AMOUNT * Math.random();
28                    bank.transfer(fromAccount, toAccount, amount);
29                    Thread.sleep((int) (DELAY * Math.random()));
30                 }
31              }
32              catch (InterruptedException e)
33              {
34              }
35           };
36           var t = new Thread(r);
37           t.start();
38        }
39     }
40  }
```

12.4.2 The Race Condition Explained

In the previous section, we ran a program in which several threads updated bank account balances. After a while, errors crept in and some amount of money was either lost or spontaneously created. This problem occurs when two threads are simultaneously trying to update an account. Suppose two threads simultaneously carry out the instruction

```
accounts[to] += amount;
```

The problem is that these are not *atomic* operations. The instruction might be processed as follows:

1. Load `accounts[to]` into a register.
2. Add `amount`.
3. Move the result back to `accounts[to]`.

Now, suppose the first thread executes Steps 1 and 2, and then it is preempted. Suppose the second thread awakens and updates the same entry in the `account` array. Then, the first thread awakens and completes its Step 3.

That action wipes out the modification of the other thread. As a result, the total is no longer correct (see Figure 12.2).

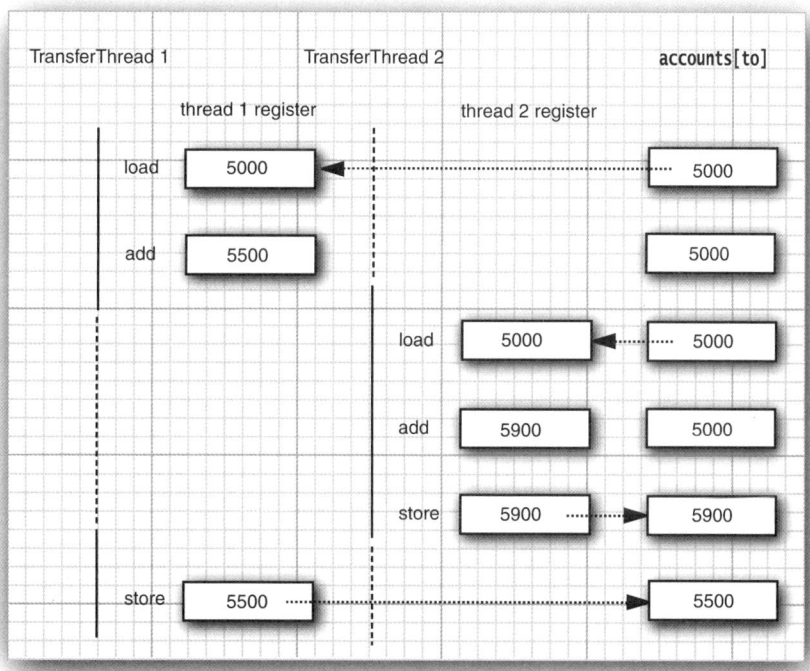

Figure 12.2 Simultaneous access by two threads

Our test program detects this corruption. (Of course, there is a slight chance of false alarms if the thread is interrupted as it is performing the tests!)

 NOTE: You can actually peek at the virtual machine bytecodes that execute each statement in our class. Run the command

```
javap -c -v Bank
```

to decompile the Bank.class file. For example, the line

```
accounts[to] += amount;
```

is translated into the following bytecodes:

```
aload_0
getfield        #2; //Field accounts:[D
iload_2
dup2
daload
dload_3
dadd
dastore
```

What these codes mean does not matter. The point is that the increment command is made up of several instructions, and the thread executing them can be interrupted at any instruction.

What is the chance of this corruption occurring? On a modern processor with multiple cores, the risk of corruption is quite high. We boosted the chance of observing the problem on a single-core processor by interleaving the print statements with the statements that update the balance.

If you omit the print statements, the risk of corruption is lower because each thread does so little work before going to sleep again, and it is unlikely that the scheduler will preempt it in the middle of the computation. However, the risk of corruption does not go away completely. If you run lots of threads on a heavily loaded machine, the program will still fail even after you have eliminated the print statements. The failure may take a few minutes or hours or days to occur. Frankly, there are few things worse in the life of a programmer than an error that only manifests itself irregularly.

The real problem is that the work of the transfer method can be interrupted in the middle. If we could ensure that the method runs to completion before the thread loses control, the state of the bank account object would never be corrupted.

12.4.3 Lock Objects

There are two mechanisms for protecting a code block from concurrent access. The Java language provides a synchronized keyword for this purpose, and Java 5 introduced the ReentrantLock class. The synchronized keyword automatically provides a lock as well as an associated "condition," which makes it powerful and convenient for most cases that require explicit locking. However, we believe that it is easier to understand the synchronized keyword after you have seen locks and conditions in isolation. The java.util.concurrent framework provides separate classes for these fundamental mechanisms, which we explain here and in Section 12.4.4, "Condition Objects," on p. 758. Once you have understood these building blocks, we present the synchronized keyword in Section 12.4.5, "The synchronized Keyword," on p. 764.

The basic outline for protecting a code block with a ReentrantLock is:

```
myLock.lock(); // a ReentrantLock object
try
{
   critical section
}
finally
{
   myLock.unlock(); // make sure the lock is unlocked even if an exception is thrown
}
```

This construct guarantees that only one thread at a time can enter the critical section. As soon as one thread locks the lock object, no other thread can get past the lock statement. When other threads call lock, they are deactivated until the first thread unlocks the lock object.

CAUTION: It is critically important that the unlock operation is enclosed in a finally clause. If the code in the critical section throws an exception, the lock must be unlocked. Otherwise, the other threads will be blocked forever.

NOTE: When you use locks, you cannot use the try-with-resources statement. First off, the unlock method isn't called close. But even if it was renamed, the try-with-resources statement wouldn't work. Its header expects the declaration of a new variable. But when you use a lock, you want to keep using the same variable that is shared among threads.

Let us use a lock to protect the transfer method of the Bank class.

```
public class Bank
{
    private var bankLock = new ReentrantLock();
    . . .
    public void transfer(int from, int to, int amount)
    {
        bankLock.lock();
        try
        {
            System.out.print(Thread.currentThread());
            accounts[from] -= amount;
            System.out.printf(" %10.2f from %d to %d", amount, from, to);
            accounts[to] += amount;
            System.out.printf(" Total Balance: %10.2f%n", getTotalBalance());
        }
        finally
        {
            bankLock.unlock();
        }
    }
}
```

Suppose one thread calls transfer and gets preempted before it is done. Suppose a second thread also calls transfer. The second thread cannot acquire the lock and is blocked in the call to the lock method. It is deactivated and must wait for the first thread to finish executing the transfer method. When the first thread unlocks the lock, then the second thread can proceed (see Figure 12.3).

Try it out. Add the locking code to the transfer method and run the program again. You can run it forever, and the bank balance will not become corrupted.

Note that each Bank object has its own ReentrantLock object. If two threads try to access the same Bank object, then the lock serves to serialize the access. However, if two threads access different Bank objects, each thread acquires a different lock and neither thread is blocked. This is as it should be, because the threads cannot interfere with one another when they manipulate different Bank instances.

The lock is called *reentrant* because a thread can repeatedly acquire a lock that it already owns. The lock has a *hold count* that keeps track of the nested calls to the lock method. The thread has to call unlock for every call to lock in order to relinquish the lock. Because of this feature, code protected by a lock can call another method that uses the same locks.

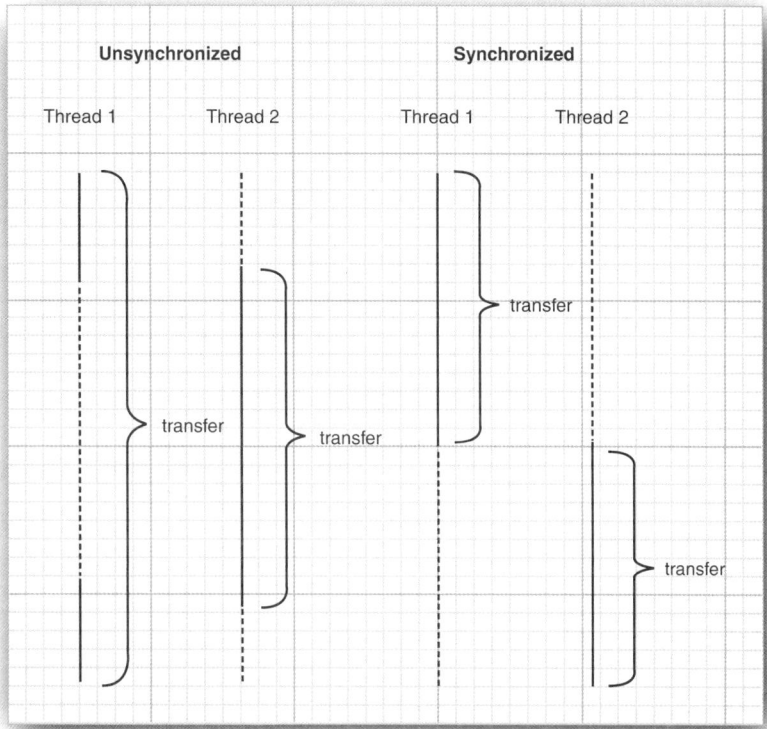

Figure 12.3 Comparison of unsynchronized and synchronized threads

For example, the transfer method calls the getTotalBalance method, which also locks the bankLock object, which now has a hold count of 2. When the getTotalBalance method exits, the hold count is back to 1. When the transfer method exits, the hold count is 0, and the thread relinquishes the lock.

In general, you will want to protect blocks of code that update or inspect a shared object, so you can be assured that these operations run to completion before another thread can use the same object.

 CAUTION: Be careful to ensure that the code in a critical section is not bypassed by throwing an exception. If an exception is thrown before the end of the section, the finally clause will relinquish the lock, but the object may be in a damaged state.

---

java.util.concurrent.locks.Lock 5

- void lock()

  acquires this lock; blocks if the lock is currently owned by another thread.

- void unlock()

  releases this lock.

---

java.util.concurrent.locks.ReentrantLock 5

- ReentrantLock()

  constructs a reentrant lock that can be used to protect a critical section.

- ReentrantLock(boolean fair)

  constructs a lock with the given fairness policy. A fair lock favors the thread that has been waiting for the longest time. However, this fairness guarantee can be a significant drag on performance. Therefore, by default, locks are not required to be fair.

CAUTION: It sounds nice to be fair, but fair locks are *a lot slower* than regular locks. You should only enable fair locking if you truly know what you are doing and have a specific reason to consider fairness essential for your program. Even if you use a fair lock, you have no guarantee that the thread scheduler is fair. If the thread scheduler chooses to neglect a thread that has been waiting a long time for the lock, it doesn't get the chance to be treated fairly by the lock.

12.4.4 Condition Objects

Often, a thread enters a critical section only to discover that it can't proceed until a condition is fulfilled. Use a *condition object* to manage threads that have acquired a lock but cannot do useful work. In this section, we introduce the implementation of condition objects in the Java library. (For historical reasons, condition objects are often called *condition variables*.)

Let us refine our simulation of the bank. We do not want to transfer money out of an account that does not have the funds to cover the transfer. Note that we cannot use code like

```
if (bank.getBalance(from) >= amount)
    bank.transfer(from, to, amount);
```

It is entirely possible that the current thread will be deactivated between the successful outcome of the test and the call to transfer.

```
if (bank.getBalance(from) >= amount)
    // thread might be deactivated at this point
  bank.transfer(from, to, amount);
```

By the time the thread is running again, the account balance may have fallen below the withdrawal amount. You must make sure that no other thread can modify the balance between the test and the transfer action. You do so by protecting both the test and the transfer action with a lock:

```
public void transfer(int from, int to, int amount)
{
   bankLock.lock();
   try
   {
      while (accounts[from] < amount)
      {
         // wait
         . . .
      }
      // transfer funds
      . . .
   }
   finally
   {
      bankLock.unlock();
   }
}
```

Now, what do we do when there is not enough money in the account? We wait until some other thread has added funds. But this thread has just gained exclusive access to the bankLock, so no other thread has a chance to make a deposit. This is where condition objects come in.

A lock object can have one or more associated condition objects. You obtain a condition object with the newCondition method. It is customary to give each condition object a name that evokes the condition that it represents. For example, here we set up a condition object to represent the "sufficient funds" condition.

```
class Bank
{
   private Condition sufficientFunds;
   . . .
   public Bank()
   {
      . . .
      sufficientFunds = bankLock.newCondition();
   }
}
```

If the transfer method finds that sufficient funds are not available, it calls

```
sufficientFunds.await();
```

The current thread is now deactivated and gives up the lock. This lets in another thread that can, we hope, increase the account balance.

There is an essential difference between a thread that is waiting to acquire a lock and a thread that has called await. Once a thread calls the await method, it enters a *wait set* for that condition. The thread is *not* made runnable when the lock is available. Instead, it stays deactivated until another thread has called the signalAll method on the same condition.

When another thread has transferred money, it should call

```
sufficientFunds.signalAll();
```

This call reactivates all threads waiting for the condition. When the threads are removed from the wait set, they are again runnable and the scheduler will eventually activate them again. At that time, they will attempt to reenter the object. As soon as the lock is available, one of them will acquire the lock *and continue where it left off*, returning from the call to await.

At this time, the thread should test the condition again. There is no guarantee that the condition is now fulfilled—the signalAll method merely signals to the waiting threads that it *may be* fulfilled at this time and that it is worth checking for the condition again.

 NOTE: In general, a call to await should be inside a loop of the form

```
while (!(OK to proceed))
    condition.await();
```

It is crucially important that *some* other thread calls the signalAll method eventually. When a thread calls await, it has no way of reactivating itself. It puts its faith in the other threads. If none of them bother to reactivate the waiting thread, it will never run again. This can lead to unpleasant *deadlock* situations. If all other threads are blocked and the last active thread calls await without unblocking one of the others, it also blocks. No thread is left to unblock the others, and the program hangs.

When should you call signalAll? The rule of thumb is to call signalAll whenever the state of an object changes in a way that might be advantageous to waiting threads. For example, whenever an account balance changes, the waiting threads should be given another chance to inspect the balance. In our example, we call `signalAll` when we have finished the funds transfer.

```
public void transfer(int from, int to, int amount)
{
   bankLock.lock();
   try
   {
      while (accounts[from] < amount)
         sufficientFunds.await();
      // transfer funds
      . . .
      sufficientFunds.signalAll();
   }
   finally
   {
      bankLock.unlock();
   }
}
```

Note that the call to `signalAll` does not immediately activate a waiting thread. It only unblocks the waiting threads so that they can compete for entry into the object after the current thread has relinquished the lock.

Another method, `signal`, unblocks only a single thread from the wait set, chosen at random. That is more efficient than unblocking all threads, but there is a danger. If the randomly chosen thread finds that it still cannot proceed, it becomes blocked again. If no other thread calls `signal` again, the system deadlocks.

CAUTION: A thread can only call `await`, `signalAll`, or `signal` on a condition if it owns the lock of the condition.

If you run the sample program in Listing 12.4, you will notice that nothing ever goes wrong. The total balance stays at $100,000 forever. No account ever has a negative balance. (Again, press Ctrl+C to terminate the program.) You may also notice that the program runs a bit slower—that is the price you pay for the added bookkeeping involved in the synchronization mechanism.

In practice, using conditions correctly can be quite challenging. Before you start implementing your own condition objects, you should consider using one of the constructs described in Section 12.5, "Thread-Safe Collections," on p. 781.

Listing 12.4 synch/Bank.java

```java
1  package synch;
2
3  import java.util.*;
4  import java.util.concurrent.locks.*;
5
6  /**
7   * A bank with a number of bank accounts that uses locks for serializing access.
8   */
9  public class Bank
10 {
11    private final double[] accounts;
12    private Lock bankLock;
13    private Condition sufficientFunds;
14
15    /**
16     * Constructs the bank.
17     * @param n the number of accounts
18     * @param initialBalance the initial balance for each account
19     */
20    public Bank(int n, double initialBalance)
21    {
22       accounts = new double[n];
23       Arrays.fill(accounts, initialBalance);
24       bankLock = new ReentrantLock();
25       sufficientFunds = bankLock.newCondition();
26    }
27
28    /**
29     * Transfers money from one account to another.
30     * @param from the account to transfer from
31     * @param to the account to transfer to
32     * @param amount the amount to transfer
33     */
34    public void transfer(int from, int to, double amount) throws InterruptedException
35    {
36       bankLock.lock();
37       try
38       {
39          while (accounts[from] < amount)
40             sufficientFunds.await();
41          System.out.print(Thread.currentThread());
42          accounts[from] -= amount;
43          System.out.printf(" %10.2f from %d to %d", amount, from, to);
44          accounts[to] += amount;
45          System.out.printf(" Total Balance: %10.2f%n", getTotalBalance());
46          sufficientFunds.signalAll();
47       }
```

```
48        finally
49        {
50           bankLock.unlock();
51        }
52     }
53
54     /**
55      * Gets the sum of all account balances.
56      * @return the total balance
57      */
58     public double getTotalBalance()
59     {
60        bankLock.lock();
61        try
62        {
63           double sum = 0;
64
65           for (double a : accounts)
66              sum += a;
67
68           return sum;
69        }
70        finally
71        {
72           bankLock.unlock();
73        }
74     }
75
76     /**
77      * Gets the number of accounts in the bank.
78      * @return the number of accounts
79      */
80     public int size()
81     {
82        return accounts.length;
83     }
84  }
```

java.util.concurrent.locks.Lock 5

- Condition newCondition()

  returns a condition object associated with this lock.

---

java.util.concurrent.locks.Condition 5

- void await()

    puts this thread on the wait set for this condition.

- void signalAll()

    unblocks all threads in the wait set for this condition.

- void signal()

    unblocks one randomly selected thread in the wait set for this condition.

12.4.5 The synchronized Keyword

In the preceding sections, you saw how to use Lock and Condition objects. Before going any further, let us summarize the key points about locks and conditions:

- A lock protects sections of code, allowing only one thread to execute the code at a time.
- A lock manages threads that are trying to enter a protected code segment.
- A lock can have one or more associated condition objects.
- Each condition object manages threads that have entered a protected code section but that cannot proceed.

The Lock and Condition interfaces give programmers a high degree of control over locking. However, in most situations, you don't need that control—you can use a mechanism that is built into the Java language. Ever since version 1.0, *every object* in Java has an intrinsic lock. If a method is declared with the synchronized keyword, the object's lock protects the entire method. That is, to call the method, a thread must acquire the intrinsic object lock.

In other words,

```
public synchronized void method()
{
    method body
}
```

is the equivalent of

```
public void method()
{
    this.intrinsicLock.lock();
    try
    {
        method body
    }
```

```
  finally { this.intrinsicLock.unlock(); }
}
```

For example, instead of using an explicit lock, we can simply declare the transfer method of the Bank class as synchronized.

The intrinsic object lock has a single associated condition. The wait method adds a thread to the wait set, and the notifyAll/notify methods unblock waiting threads. In other words, calling wait or notifyAll is the equivalent of

```
intrinsicCondition.await();
intrinsicCondition.signalAll();
```

NOTE: The wait, notifyAll, and notify methods are final methods of the Object class. The Condition methods had to be named await, signalAll, and signal so that they don't conflict with those methods.

For example, you can implement the Bank class in Java like this:

```
class Bank
{
   private double[] accounts;

   public synchronized void transfer(int from, int to, int amount)
         throws InterruptedException
   {
      while (accounts[from] < amount)
         wait(); // wait on intrinsic object lock's single condition
      accounts[from] -= amount;
      accounts[to] += amount;
      notifyAll(); // notify all threads waiting on the condition
   }

   public synchronized double getTotalBalance() { . . . }
}
```

As you can see, using the synchronized keyword yields code that is much more concise. Of course, to understand this code, you have to know that each object has an intrinsic lock, and that the lock has an intrinsic condition. The lock manages the threads that try to enter a synchronized method. The condition manages the threads that have called wait.

TIP: Synchronized methods are relatively straightforward. However, beginners often struggle with conditions. Before you use wait/notifyAll, you should consider using one of the constructs described in Section 12.5, "Thread-Safe Collections," on p. 781.

It is also legal to declare static methods as synchronized. If such a method is called, it acquires the intrinsic lock of the associated class object. For example, if the Bank class has a static synchronized method, then the lock of the Bank.class object is locked when it is called. As a result, no other thread can call this or any other synchronized static method of the same class.

The intrinsic locks and conditions have some limitations. Among them:

- You cannot interrupt a thread that is trying to acquire a lock.
- You cannot specify a timeout when trying to acquire a lock.
- Having a single condition per lock can be inefficient.

What should you use in your code—Lock and Condition objects or synchronized methods? Here is our recommendation:

- It is best to use neither Lock/Condition nor the synchronized keyword. In many situations, you can use one of the mechanisms of the java.util.concurrent package that do all the locking for you. For example, in Section 12.5.1, "Blocking Queues," on p. 781, you will see how to use a blocking queue to synchronize threads that work on a common task. You should also explore parallel streams—see Chapter 1 of Volume II.
- If the synchronized keyword works for your situation, by all means, use it. You'll write less code and have less room for error. Listing 12.5 shows the bank example, implemented with synchronized methods.
- Use Lock/Condition if you really need the additional power that these constructs give you.

Listing 12.5 synch2/Bank.java

```
1  package synch2;
2
3  import java.util.*;
4
5  /**
6   * A bank with a number of bank accounts that uses synchronization primitives.
7   */
8  public class Bank
9  {
10     private final double[] accounts;
11
12     /**
13      * Constructs the bank.
14      * @param n the number of accounts
15      * @param initialBalance the initial balance for each account
16      */
```

```
17    public Bank(int n, double initialBalance)
18    {
19       accounts = new double[n];
20       Arrays.fill(accounts, initialBalance);
21    }
22
23    /**
24     * Transfers money from one account to another.
25     * @param from the account to transfer from
26     * @param to the account to transfer to
27     * @param amount the amount to transfer
28     */
29    public synchronized void transfer(int from, int to, double amount)
30          throws InterruptedException
31    {
32       while (accounts[from] < amount)
33          wait();
34       System.out.print(Thread.currentThread());
35       accounts[from] -= amount;
36       System.out.printf(" %10.2f from %d to %d", amount, from, to);
37       accounts[to] += amount;
38       System.out.printf(" Total Balance: %10.2f%n", getTotalBalance());
39       notifyAll();
40    }
41
42    /**
43     * Gets the sum of all account balances.
44     * @return the total balance
45     */
46    public synchronized double getTotalBalance()
47    {
48       double sum = 0;
49
50       for (double a : accounts)
51          sum += a;
52
53       return sum;
54    }
55
56    /**
57     * Gets the number of accounts in the bank.
58     * @return the number of accounts
59     */
60    public int size()
61    {
62       return accounts.length;
63    }
64 }
```

---

java.lang.Object 1.0

- void notifyAll()

  unblocks the threads that called wait on this object. This method can only be called from within a synchronized method or block. The method throws an IllegalMonitorStateException if the current thread is not the owner of the object's lock.

- void notify()

  unblocks one randomly selected thread among the threads that called wait on this object. This method can only be called from within a synchronized method or block. The method throws an IllegalMonitorStateException if the current thread is not the owner of the object's lock.

- void wait()

  causes a thread to wait until it is notified. This method can only be called from within a synchronized method or block. It throws an IllegalMonitorStateException if the current thread is not the owner of the object's lock.

- void wait(long millis)
- void wait(long millis, int nanos)

  causes a thread to wait until it is notified or until the specified amount of time has passed. These methods can only be called from within a synchronized method or block. They throw an IllegalMonitorStateException if the current thread is not the owner of the object's lock. The number of nanoseconds may not exceed 1,000,000.

12.4.6 Synchronized Blocks

As we just discussed, every Java object has a lock. A thread can acquire the lock by calling a synchronized method. There is a second mechanism for acquiring the lock: by entering a *synchronized block*. When a thread enters a block of the form

```
synchronized (obj) // this is the syntax for a synchronized block
{
   critical section
}
```

then it acquires the lock for obj.

You will sometimes find "ad hoc" locks, such as

```
public class Bank
{
   private double[] accounts;
   private var lock = new Object();
```

```
. . .
public void transfer(int from, int to, int amount)
{
   synchronized (lock) // an ad-hoc lock
   {
      accounts[from] -= amount;
      accounts[to] += amount;
   }
   System.out.println(. . .);
}
}
```

Here, the lock object is created only to use the lock that every Java object possesses.

Sometimes, programmers use the lock of an object to implement additional atomic operations—a practice known as *client-side locking*. Consider, for example, the Vector class, which is a list whose methods are synchronized. Now suppose we stored our bank balances in a Vector<Double>. Here is a naive implementation of a transfer method:

```
public void transfer(Vector<Double> accounts, int from, int to, int amount) // ERROR
{
   accounts.set(from, accounts.get(from) - amount);
   accounts.set(to, accounts.get(to) + amount);
   System.out.println(. . .);
}
```

The get and set methods of the Vector class are synchronized, but that doesn't help us. It is entirely possible for a thread to be preempted in the transfer method after the first call to get has been completed. Another thread may then store a different value into the same position. However, we can hijack the lock:

```
public void transfer(Vector<Double> accounts, int from, int to, int amount)
{
   synchronized (accounts)
   {
      accounts.set(from, accounts.get(from) - amount);
      accounts.set(to, accounts.get(to) + amount);
   }
   System.out.println(. . .);
}
```

This approach works, but it is entirely dependent on the fact that the Vector class uses the intrinsic lock for all of its mutator methods. However, is this really a fact? The documentation of the Vector class makes no such promise. You have to carefully study the source code and hope that future versions do not introduce unsynchronized mutators. As you can see, client-side locking is very fragile and not generally recommended.

 NOTE: The Java virtual machine has built-in support for synchronized methods. However, synchronized blocks are compiled into a lengthy sequence of bytecodes to manage the intrinsic lock.

12.4.7 The Monitor Concept

Locks and conditions are powerful tools for thread synchronization, but they are not very object-oriented. For many years, researchers have looked for ways to make multithreading safe without forcing programmers to think about explicit locks. One of the most successful solutions is the *monitor* concept that was pioneered by Per Brinch Hansen and Tony Hoare in the 1970s. In the terminology of Java, a monitor has these properties:

- A monitor is a class with only private fields.
- Each object of that class has an associated lock.
- All methods are locked by that lock. In other words, if a client calls obj.method(), then the lock for obj is automatically acquired at the beginning of the method call and relinquished when the method returns. Since all fields are private, this arrangement ensures that no thread can access the fields while another thread manipulates them.
- The lock can have any number of associated conditions.

Earlier versions of monitors had a single condition, with a rather elegant syntax. You can simply call await accounts[from] >= amount without using an explicit condition variable. However, research showed that indiscriminate retesting of conditions can be inefficient. This problem is solved with explicit condition variables, each managing a separate set of threads.

The Java designers loosely adapted the monitor concept. *Every object* in Java has an intrinsic lock and an intrinsic condition. If a method is declared with the synchronized keyword, it acts like a monitor method. The condition variable is accessed by calling wait/notifyAll/notify.

However, a Java object differs from a monitor in three important ways, compromising thread safety:

- Fields are not required to be private.
- Methods are not required to be synchronized.
- The intrinsic lock is available to clients.

This disrespect for security enraged Per Brinch Hansen. In a scathing review of the multithreading primitives in Java, he wrote: "It is astounding to me that Java's insecure parallelism is taken seriously by the programming community, a quarter of a century after the invention of monitors and Concurrent Pascal. It has no merit" [Java's Insecure Parallelism, *ACM SIGPLAN Notices* 34:38–45, April 1999].

12.4.8 Volatile Fields

Sometimes, it seems excessive to pay the cost of synchronization just to read or write an instance field or two. After all, what can go wrong? Unfortunately, with modern processors and compilers, there is plenty of room for error.

- Computers with multiple processors can temporarily hold memory values in registers or local memory caches. As a consequence, threads running in different processors may see different values for the same memory location!

- Compilers can reorder instructions for maximum throughput. Compilers won't choose an ordering that changes the meaning of the code, but they make the assumption that memory values are only changed when there are explicit instructions in the code. However, a memory value can be changed by another thread!

If you use locks to protect code that can be accessed by multiple threads, you won't have these problems. Compilers are required to respect locks by flushing local caches as necessary and not inappropriately reordering instructions. The details are explained in the Java Memory Model and Thread Specification developed by JSR 133 (see www.jcp.org/en/jsr/detail?id=133). Much of the specification is highly complex and technical, but the document also contains a number of clearly explained examples. A more accessible overview article by Brian Goetz is available at www.ibm.com/developerworks/library/j-jtp02244.

 NOTE: Brian Goetz coined the following "synchronization motto": "If you write a variable which may next be read by another thread, or you read a variable which may have last been written by another thread, you must use synchronization."

The volatile keyword offers a lock-free mechanism for synchronizing access to an instance field. If you declare a field as volatile, then the compiler and the virtual machine take into account that the field may be concurrently updated by another thread.

For example, suppose an object has a `boolean` flag `done` that is set by one thread and queried by another thread. As we already discussed, you can use a lock:

```
private boolean done;
public synchronized boolean isDone() { return done; }
public synchronized void setDone() { done = true; }
```

Perhaps it is not a good idea to use the intrinsic object lock. The `isDone` and `setDone` methods can block if another thread has locked the object. If that is a concern, one can use a separate lock just for this variable. But this is getting to be a lot of trouble.

In this case, it is reasonable to declare the field as `volatile`:

```
private volatile boolean done;
public boolean isDone() { return done; }
public void setDone() { done = true; }
```

The compiler will insert the appropriate code to ensure that a change to the `done` variable in one thread is visible from any other thread that reads the variable.

 CAUTION: Volatile variables do not provide any atomicity. For example, the method

```
public void flipDone() { done = !done; } // not atomic
```

is not guaranteed to flip the value of the field. There is no guarantee that the reading, flipping, and writing is uninterrupted.

12.4.9 Final Variables

As you saw in the preceding section, you cannot safely read a field from multiple threads unless you use locks or the `volatile` modifier.

There is one other situation in which it is safe to access a shared field—when it is declared `final`. Consider

```
final var accounts = new HashMap<String, Double>();
```

Other threads get to see the `accounts` variable after the constructor has finished.

Without using `final`, there would be no guarantee that other threads would see the updated value of `accounts`—they might all see `null`, not the constructed `HashMap`.

Of course, the operations on the map are not thread-safe. If multiple threads mutate and read the map, you still need synchronization.

12.4.10 Atomics

You can declare shared variables as volatile provided you perform no operations other than assignment.

There are a number of classes in the java.util.concurrent.atomic package that use efficient machine-level instructions to guarantee atomicity of other operations without using locks. For example, the AtomicInteger class has methods incrementAndGet and decrementAndGet that atomically increment or decrement an integer. For example, you can safely generate a sequence of numbers like this:

```
public static AtomicLong nextNumber = new AtomicLong();
// in some thread. . .
long id = nextNumber.incrementAndGet();
```

The incrementAndGet method atomically increments the AtomicLong and returns the post-increment value. That is, the operations of getting the value, adding 1, setting it, and producing the new value cannot be interrupted. It is guaranteed that the correct value is computed and returned, even if multiple threads access the same instance concurrently.

There are methods for atomically setting, adding, and subtracting values, but if you want to make a more complex update, you have to use the compareAndSet method. For example, suppose you want to keep track of the largest value that is observed by different threads. The following won't work:

```
public static AtomicLong largest = new AtomicLong();
// in some thread. . .
largest.set(Math.max(largest.get(), observed)); // ERROR--race condition!
```

This update is not atomic. Instead, provide a lambda expression for updating the variable, and the update is done for you. In our example, we can call

```
largest.updateAndGet(x -> Math.max(x, observed));
```

or

```
largest.accumulateAndGet(observed, Math::max);
```

The accumulateAndGet method takes a binary operator that is used to combine the atomic value and the supplied argument.

There are also methods getAndUpdate and getAndAccumulate that return the old value.

 NOTE: These methods are also provided for the classes AtomicInteger, AtomicIntegerArray, AtomicIntegerFieldUpdater, AtomicLongArray, AtomicLongFieldUpdater, AtomicReference, AtomicReferenceArray, and AtomicReferenceFieldUpdater.

When you have a very large number of threads accessing the same atomic values, performance suffers because the optimistic updates require too many retries. The LongAdder and LongAccumulator classes solve this problem. A LongAdder is composed of multiple variables whose collective sum is the current value. Multiple threads can update different summands, and new summands are automatically provided when the number of threads increases. This is efficient in the common situation where the value of the sum is not needed until after all work has been done. The performance improvement can be substantial.

If you anticipate high contention, you should simply use a LongAdder instead of an AtomicLong. The method names are slightly different. Call increment to increment a counter or add to add a quantity, and sum to retrieve the total.

```
var adder = new LongAdder();
for (. . .)
   pool.submit(() -> {
      while (. . .) {
         . . .
         if (. . .) adder.increment();
      }
   });
. . .
long total = adder.sum();
```

 NOTE: Of course, the increment method does *not* return the old value. Doing that would undo the efficiency gain of splitting the sum into multiple summands.

The LongAccumulator generalizes this idea to an arbitrary accumulation operation. In the constructor, you provide the operation, as well as its neutral element. To incorporate new values, call accumulate. Call get to obtain the current value. The following has the same effect as a LongAdder:

```
var adder = new LongAccumulator(Long::sum, 0);
// in some thread. . .
adder.accumulate(value);
```

Internally, the accumulator has variables a_1, a_2, . . ., a_n. Each variable is initialized with the neutral element (0 in our example).

When accumulate is called with value v, then one of them is atomically updated as $a_i = a_i\ op\ v$, where op is the accumulation operation written in infix form. In our example, a call to accumulate computes $a_i = a_i + v$ for some i.

The result of get is $a_1\ op\ a_2\ op\ \ldots\ op\ a_n$. In our example, that is the sum of the accumulators, $a_1 + a_2 + \ldots + a_n$.

If you choose a different operation, you can compute maximum or minimum. In general, the operation must be associative and commutative. That means that the final result must be independent of the order in which the intermediate values were combined.

There are also `DoubleAdder` and `DoubleAccumulator` that work in the same way, except with `double` values.

12.4.11 Deadlocks

Locks and conditions cannot solve all problems that might arise in multithreading. Consider the following situation:

1. Account 1: $200
2. Account 2: $300
3. Thread 1: Transfer $300 from Account 1 to Account 2
4. Thread 2: Transfer $400 from Account 2 to Account 1

As Figure 12.4 indicates, Threads 1 and 2 are clearly blocked. Neither can proceed because the balances in Accounts 1 and 2 are insufficient.

It is possible that all threads get blocked because each is waiting for more money. Such a situation is called a *deadlock*.

In our program, a deadlock cannot occur for a simple reason. Each transfer amount is for, at most, $1,000. Since there are 100 accounts and a total of $100,000 in them, at least one of the accounts must have at least $1,000 at any time. The thread moving money out of that account can therefore proceed.

But if you change the `run` method of the threads to remove the $1,000 transaction limit, deadlocks can occur quickly. Try it out. Set `NACCOUNTS` to 10. Construct each transfer runnable with a `max` value of 2 * `INITIAL_BALANCE` and run the program. The program will run for a while and then hang.

 TIP: When the program hangs, press Ctrl+\. You will get a thread dump that lists all threads. Each thread has a stack trace, telling you where it is currently blocked. Alternatively, run `jconsole`, as described in Chapter 7, and consult the Threads panel (see Figure 12.5).

Another way to create a deadlock is to make the ith thread responsible for putting money into the ith account, rather than for taking it out of the ith account. In this case, there is a chance that all threads will gang up on one account, each trying to remove more money from it than it contains. Try it out. In the `SynchBankTest` program, turn to the `run` method of the `TransferRunnable`

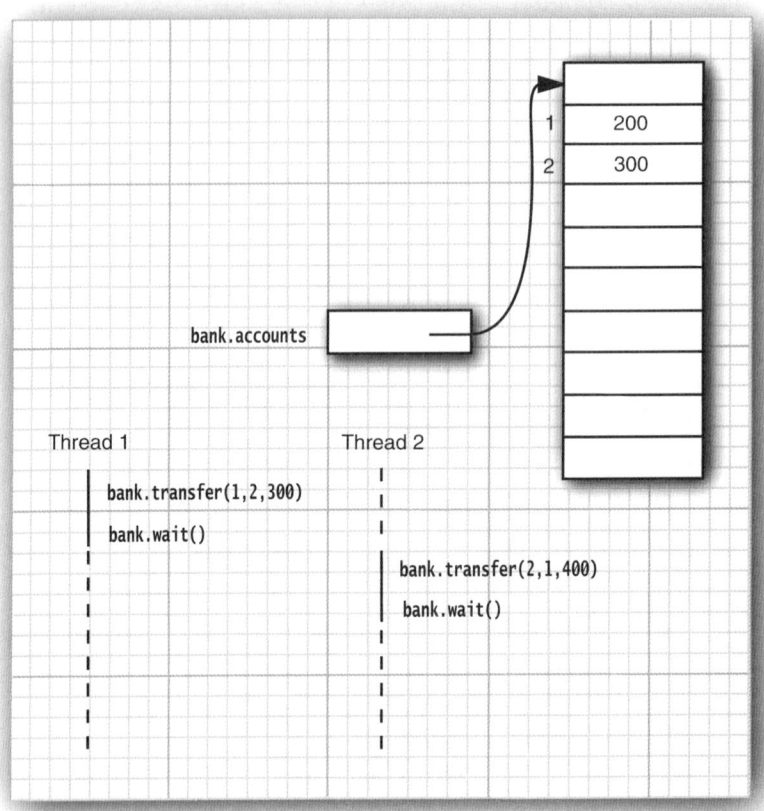

Figure 12.4 A deadlock situation

class. In the call to transfer, flip fromAccount and toAccount. Run the program and see how it deadlocks almost immediately.

Here is another situation in which a deadlock can occur easily. Change the signalAll method to signal in the SynchBankTest program. You will find that the program eventually hangs. (Again, set NACCOUNTS to 10 to observe the effect more quickly.) Unlike signalAll, which notifies all threads that are waiting for added funds, the signal method unblocks only one thread. If that thread can't proceed, all threads can be blocked. Consider the following sample scenario of a developing deadlock:

1. Account 1: $1,990

2. All other accounts: $990 each

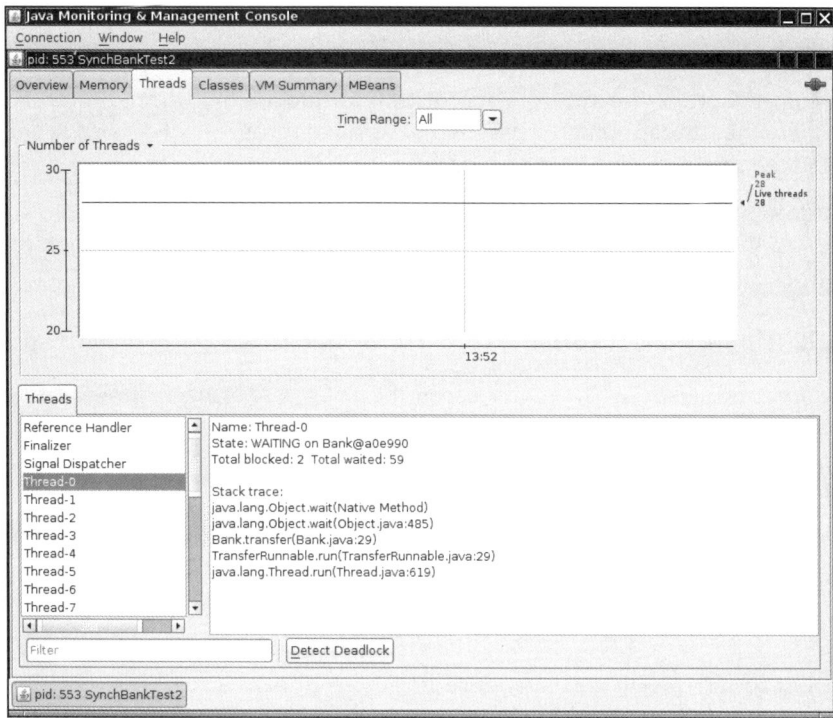

Figure 12.5 The Threads panel in jconsole

3. Thread 1: Transfer $995 from Account 1 to Account 2

4. All other threads: Transfer $995 from their account to another account

Clearly, all threads but Thread 1 are blocked, because there isn't enough money in their accounts.

Thread 1 proceeds. Afterward, we have the following situation:

1. Account 1: $995

2. Account 2: $1,985

3. All other accounts: $990 each

Then, Thread 1 calls signal. The signal method picks a thread at random to unblock. Suppose it picks Thread 3. That thread is awakened, finds that there isn't enough money in its account, and calls await again. But Thread 1 is still running. A new random transaction is generated, say,

1. Thread 1: Transfer $997 from Account 1 to Account 2

Now, Thread 1 also calls await, and *all* threads are blocked. The system has deadlocked.

The culprit here is the call to signal. It only unblocks one thread, and it may not pick the thread that is essential to make progress. (In our scenario, Thread 2 must proceed to take money out of Account 2.)

Unfortunately, there is nothing in the Java programming language to avoid or break these deadlocks. You must design your program to ensure that a deadlock situation cannot occur.

12.4.12 Thread-Local Variables

In the preceding sections, we discussed the risks of sharing variables between threads. Sometimes, you can avoid sharing by giving each thread its own instance, using the ThreadLocal helper class. For example, the SimpleDateFormat class is not thread-safe. Suppose we have a static variable

```
public static final SimpleDateFormat dateFormat = new SimpleDateFormat("yyyy-MM-dd");
```

If two threads execute an operation such as

```
String dateStamp = dateFormat.format(new Date());
```

then the result can be garbage since the internal data structures used by the dateFormat can be corrupted by concurrent access. You could use synchronization, which is expensive, or you could construct a local SimpleDateFormat object whenever you need it, but that is also wasteful.

To construct one instance per thread, use the following code:

```
public static final ThreadLocal<SimpleDateFormat> dateFormat
   = ThreadLocal.withInitial(() -> new SimpleDateFormat("yyyy-MM-dd"));
```

To access the actual formatter, call

```
String dateStamp = dateFormat.get().format(new Date());
```

The first time you call get in a given thread, the lambda in the constructor is called. From then on, the get method returns the instance belonging to the current thread.

A similar problem is the generation of random numbers in multiple threads. The java.util.Random class is thread-safe. But it is still inefficient if multiple threads need to wait for a single shared generator.

You could use the ThreadLocal helper to give each thread a separate generator, but Java 7 provides a convenience class for you. Simply make a call such as

```
int random = ThreadLocalRandom.current().nextInt(upperBound);
```

The call ThreadLocalRandom.current() returns an instance of the Random class that is unique to the current thread.

---

java.lang.ThreadLocal<T> 1.2

- T get()

  gets the current value of this thread. If get is called for the first time, the value is obtained by calling initialize.

- void set(T t)

  sets a new value for this thread.

- void remove()

  removes the value for this thread.

- static <S> ThreadLocal<S> withInitial(Supplier<? extends S> supplier) 8

  creates a thread local variable whose initial value is produced by invoking the given supplier.

---

java.util.concurrent.ThreadLocalRandom 7

- static ThreadLocalRandom current()

  returns an instance of the Random class that is unique to the current thread.

12.4.13 Why the stop and suspend Methods Are Deprecated

The initial release of Java defined a stop method that simply terminates a thread, and a suspend method that blocks a thread until another thread calls resume. The stop and suspend methods have something in common: Both attempt to control the behavior of a given thread without the thread's cooperation.

The stop, suspend, and resume methods have been deprecated. The stop method is inherently unsafe, and experience has shown that the suspend method frequently leads to deadlocks. In this section, you will see why these methods are problematic and what you can do to avoid problems.

Let us turn to the stop method first. This method terminates all pending methods, including the run method. When a thread is stopped, it immediately gives up the locks on all objects that it has locked. This can leave objects in an inconsistent state. For example, suppose a TransferRunnable is stopped in the middle of moving money from one account to another, after the withdrawal and before the deposit. Now the bank object is *damaged*. Since the lock has been relinquished, the damage is observable from the other threads that have not been stopped.

When a thread wants to stop another thread, it has no way of knowing when the stop method is safe and when it leads to damaged objects. Therefore, the method has been deprecated. You should interrupt a thread when you want it to stop. The interrupted thread can then stop when it is safe to do so.

 NOTE: Some authors claim that the stop method has been deprecated because it can cause objects to be permanently locked by a stopped thread. However, that claim is not valid. A stopped thread exits all synchronized methods it has called—technically, by throwing a ThreadDeath exception. As a consequence, the thread relinquishes the intrinsic object locks that it holds.

Next, let us see what is wrong with the suspend method. Unlike stop, suspend won't damage objects. However, if you suspend a thread that owns a lock, then the lock is unavailable until the thread is resumed. If the thread that calls the suspend method tries to acquire the same lock, the program deadlocks: The suspended thread waits to be resumed, and the suspending thread waits for the lock.

This situation occurs frequently in graphical user interfaces. Suppose we have a graphical simulation of our bank. A button labeled Pause suspends the transfer threads, and a button labeled Resume resumes them.

```
pauseButton.addActionListener(event -> {
    for (int i = 0; i < threads.length; i++)
        threads[i].suspend(); // don't do this
});
resumeButton.addActionListener(event -> {
    for (int i = 0; i < threads.length; i++)
        threads[i].resume();
});
```

Suppose a paintComponent method paints a chart of each account, calling a getBalances method to get an array of balances.

As you will see in Section 12.7.3, "Long-Running Tasks in User Interface Callbacks," on p. 823, both the button actions and the repainting occur in the same thread, the *event dispatch thread*. Consider the following scenario:

1. One of the transfer threads acquires the lock of the bank object.
2. The user clicks the Pause button.
3. All transfer threads are suspended; one of them still holds the lock on the bank object.

4. For some reason, the account chart needs to be repainted.

5. The paintComponent method calls the getBalances method.

6. That method tries to acquire the lock of the bank object.

Now the program is frozen.

The event dispatch thread can't proceed because the lock is owned by one of the suspended threads. Thus, the user can't click the Resume button, and the threads won't ever resume.

If you want to safely suspend a thread, introduce a variable suspendRequested and test it in a safe place of your run method—in a place where your thread doesn't lock objects that other threads need. When your thread finds that the suspendRequested variable has been set, it should keep waiting until it becomes available again.

12.5 Thread-Safe Collections

If multiple threads concurrently modify a data structure, such as a hash table, it is easy to damage that data structure. (See Chapter 9 for more information on hash tables.) For example, one thread may begin to insert a new element. Suppose it is preempted in the middle of rerouting the links between the hash table's buckets. If another thread starts traversing the same list, it may follow invalid links and create havoc, perhaps throwing exceptions or getting trapped in an infinite loop.

You can protect a shared data structure by supplying a lock, but it is usually easier to choose a thread-safe implementation instead. In the following sections, we discuss the other thread-safe collections that the Java library provides.

12.5.1 Blocking Queues

Many threading problems can be formulated elegantly and safely by using one or more queues. Producer threads insert items into the queue, and consumer threads retrieve them. The queue lets you safely hand over data from one thread to another. For example, consider our bank transfer program. Instead of accessing the bank object directly, the transfer threads insert transfer instruction objects into a queue. Another thread removes the instructions from the queue and carries out the transfers. Only that thread has access to the internals of the bank object. No synchronization is necessary. (Of course, the implementors of the thread-safe queue classes had to worry about locks and conditions, but that was their problem, not yours.)

A *blocking queue* causes a thread to block when you try to add an element when the queue is currently full or to remove an element when the queue is empty. Blocking queues are a useful tool for coordinating the work of multiple threads. Worker threads can periodically deposit intermediate results into a blocking queue. Other worker threads remove the intermediate results and modify them further. The queue automatically balances the workload. If the first set of threads runs slower than the second, the second set blocks while waiting for the results. If the first set of threads runs faster, the queue fills up until the second set catches up. Table 12.1 shows the methods for blocking queues.

The blocking queue methods fall into three categories that differ by the action they perform when the queue is full or empty. If you use the queue as a thread management tool, use the put and take methods. The add, remove, and element operations throw an exception when you try to add to a full queue or get the head of an empty queue. Of course, in a multithreaded program, the queue might become full or empty at any time, so you will instead want to use the offer, poll, and peek methods. These methods simply return with a failure indicator instead of throwing an exception if they cannot carry out their tasks.

 NOTE: The poll and peek methods return null to indicate failure. Therefore, it is illegal to insert null values into these queues.

There are also variants of the offer and poll methods with a timeout. For example, the call

```
boolean success = q.offer(x, 100, TimeUnit.MILLISECONDS);
```

tries for 100 milliseconds to insert an element to the tail of the queue. If it succeeds, it returns true; otherwise, it returns false when it times out. Similarly, the call

```
Object head = q.poll(100, TimeUnit.MILLISECONDS);
```

tries for 100 milliseconds to remove the head of the queue. If it succeeds, it returns the head; otherwise, it returns null when it times out.

The put method blocks if the queue is full, and the take method blocks if the queue is empty. These are the equivalents of offer and poll with no timeout.

The java.util.concurrent package supplies several variations of blocking queues. By default, the LinkedBlockingQueue has no upper bound on its capacity, but a maximum capacity can be optionally specified. The LinkedBlockingDeque is a double-ended version. The ArrayBlockingQueue is constructed with a given capacity and

Table 12.1 Blocking Queue Methods

Method	Normal Action	Action in Special Circumstances
add	Adds an element	Throws an IllegalStateException if the queue is full
element	Returns the head element	Throws a NoSuchElementException if the queue is empty
offer	Adds an element and returns true	Returns false if the queue is full
peek	Returns the head element	Returns null if the queue is empty
poll	Removes and returns the head element	Returns null if the queue is empty
put	Adds an element	Blocks if the queue is full
remove	Removes and returns the head element	Throws a NoSuchElementException if the queue is empty
take	Removes and returns the head element	Blocks if the queue is empty

an optional parameter to require fairness. If fairness is specified, then the longest-waiting threads are given preferential treatment. As always, fairness exacts a significant performance penalty, and you should only use it if your problem specifically requires it.

The PriorityBlockingQueue is a priority queue, not a first-in/first-out queue. Elements are removed in order of their priority. The queue has unbounded capacity, but retrieval will block if the queue is empty. (See Chapter 9 for more information on priority queues.)

A DelayQueue contains objects that implement the Delayed interface:

```
interface Delayed extends Comparable<Delayed>
{
   long getDelay(TimeUnit unit);
}
```

The getDelay method returns the remaining delay of the object. A negative value indicates that the delay has elapsed. Elements can only be removed from a DelayQueue if their delay has elapsed. You also need to implement the compareTo method. The DelayQueue uses that method to sort the entries.

Java 7 adds a TransferQueue interface that allows a producer thread to wait until a consumer is ready to take on an item. When a producer calls

```
q.transfer(item);
```

the call blocks until another thread removes it. The LinkedTransferQueue class implements this interface.

The program in Listing 12.6 shows how to use a blocking queue to control a set of threads. The program searches through all files in a directory and its subdirectories, printing lines that contain a given keyword.

A producer thread enumerates all files in all subdirectories and places them in a blocking queue. This operation is fast, and the queue would quickly fill up with all files in the file system if it was not bounded.

We also start a large number of search threads. Each search thread takes a file from the queue, opens it, prints all lines containing the keyword, and then takes the next file. We use a trick to terminate the application when no further work is required. In order to signal completion, the enumeration thread places a dummy object into the queue. (This is similar to a dummy suitcase with a label "last bag" in a baggage claim belt.) When a search thread takes the dummy, it puts it back and terminates.

Note that no explicit thread synchronization is required. In this application, we use the queue data structure as a synchronization mechanism.

Listing 12.6 blockingQueue/BlockingQueueTest.java

```
1  package blockingQueue;
2
3  import java.io.*;
4  import java.nio.charset.*;
5  import java.nio.file.*;
6  import java.util.*;
7  import java.util.concurrent.*;
8  import java.util.stream.*;
9
10  /**
11   * @version 1.03 2018-03-17
12   * @author Cay Horstmann
13   */
14  public class BlockingQueueTest
15  {
16     private static final int FILE_QUEUE_SIZE = 10;
17     private static final int SEARCH_THREADS = 100;
18     private static final Path DUMMY = Path.of("");
19     private static BlockingQueue<Path> queue = new ArrayBlockingQueue<>(FILE_QUEUE_SIZE);
20
21     public static void main(String[] args)
22     {
```

```
23        try (var in = new Scanner(System.in))
24        {
25           System.out.print("Enter base directory (e.g. /opt/jdk-9-src): ");
26           String directory = in.nextLine();
27           System.out.print("Enter keyword (e.g. volatile): ");
28           String keyword = in.nextLine();
29
30           Runnable enumerator = () -> {
31              try
32              {
33                 enumerate(Path.of(directory));
34                 queue.put(DUMMY);
35              }
36              catch (IOException e)
37              {
38                 e.printStackTrace();
39              }
40              catch (InterruptedException e)
41              {
42              }
43           };
44
45           new Thread(enumerator).start();
46           for (int i = 1; i <= SEARCH_THREADS; i++) {
47              Runnable searcher = () -> {
48                 try
49                 {
50                    var done = false;
51                    while (!done)
52                    {
53                       Path file = queue.take();
54                       if (file == DUMMY)
55                       {
56                          queue.put(file);
57                          done = true;
58                       }
59                       else search(file, keyword);
60                    }
61                 }
62                 catch (IOException e)
63                 {
64                    e.printStackTrace();
65                 }
66                 catch (InterruptedException e)
67                 {
68                 }
69              };
```

(Continues)

Listing 12.6 *(Continued)*

```
70              new Thread(searcher).start();
71          }
72       }
73    }
74
75    /**
76     * Recursively enumerates all files in a given directory and its subdirectories.
77     * See Chapters 1 and 2 of Volume II for the stream and file operations.
78     * @param directory the directory in which to start
79     */
80    public static void enumerate(Path directory) throws IOException, InterruptedException
81    {
82       try (Stream<Path> children = Files.list(directory))
83       {
84          for (Path child : children.collect(Collectors.toList()))
85          {
86             if (Files.isDirectory(child))
87                enumerate(child);
88             else
89                queue.put(child);
90          }
91       }
92    }
93
94    /**
95     * Searches a file for a given keyword and prints all matching lines.
96     * @param file the file to search
97     * @param keyword the keyword to search for
98     */
99    public static void search(Path file, String keyword) throws IOException
100   {
101      try (var in = new Scanner(file, StandardCharsets.UTF_8))
102      {
103         int lineNumber = 0;
104         while (in.hasNextLine())
105         {
106            lineNumber++;
107            String line = in.nextLine();
108            if (line.contains(keyword))
109               System.out.printf("%s:%d:%s%n", file, lineNumber, line);
110         }
111      }
112   }
113 }
```

`java.util.concurrent.ArrayBlockingQueue<E>` 5

- `ArrayBlockingQueue(int capacity)`
- `ArrayBlockingQueue(int capacity, boolean fair)`

constructs a blocking queue with the given capacity and fairness settings. The queue is implemented as a circular array.

`java.util.concurrent.LinkedBlockingQueue<E>` 5
`java.util.concurrent.LinkedBlockingDeque<E>` 6

- `LinkedBlockingQueue()`
- `LinkedBlockingDeque()`

constructs an unbounded blocking queue or deque, implemented as a linked list.

- `LinkedBlockingQueue(int capacity)`
- `LinkedBlockingDeque(int capacity)`

constructs a bounded blocking queue or deque with the given capacity, implemented as a linked list.

`java.util.concurrent.DelayQueue<E extends Delayed>` 5

- `DelayQueue()`

constructs an unbounded blocking queue of `Delayed` elements. Only elements whose delay has expired can be removed from the queue.

`java.util.concurrent.Delayed` 5

- `long getDelay(TimeUnit unit)`

gets the delay for this object, measured in the given time unit.

`java.util.concurrent.PriorityBlockingQueue<E>` 5

- `PriorityBlockingQueue()`
- `PriorityBlockingQueue(int initialCapacity)`
- `PriorityBlockingQueue(int initialCapacity, Comparator<? super E> comparator)`

constructs an unbounded blocking priority queue implemented as a heap. The default for the initial capacity is 11. If the comparator is not specified, the elements must implement the `Comparable` interface.

---

java.util.concurrent.BlockingQueue<E>   5

- void put(E element)

  adds the element, blocking if necessary.

- E take()

  removes and returns the head element, blocking if necessary.

- boolean offer(E element, long time, TimeUnit unit)

  adds the given element and returns true if successful, blocking if necessary until the element has been added or the time has elapsed.

- E poll(long time, TimeUnit unit)

  removes and returns the head element, blocking if necessary until an element is available or the time has elapsed. Returns null upon failure.

---

java.util.concurrent.BlockingDeque<E>   6

- void putFirst(E element)
- void putLast(E element)

  adds the element, blocking if necessary.

- E takeFirst()
- E takeLast()

  removes and returns the head or tail element, blocking if necessary.

- boolean offerFirst(E element, long time, TimeUnit unit)
- boolean offerLast(E element, long time, TimeUnit unit)

  adds the given element and returns true if successful, blocking if necessary until the element has been added or the time has elapsed.

- E pollFirst(long time, TimeUnit unit)
- E pollLast(long time, TimeUnit unit)

  removes and returns the head or tail element, blocking if necessary until an element is available or the time has elapsed. Returns null upon failure.

---

java.util.concurrent.TransferQueue<E>   7

- void transfer(E element)
- boolean tryTransfer(E element, long time, TimeUnit unit)

  transfers a value, or tries transferring it with a given timeout, blocking until another thread has removed the item. The second method returns true if successful.

12.5.2 Efficient Maps, Sets, and Queues

The `java.util.concurrent` package supplies efficient implementations for maps, sorted sets, and queues: `ConcurrentHashMap`, `ConcurrentSkipListMap`, `ConcurrentSkipListSet`, and `ConcurrentLinkedQueue`.

These collections use sophisticated algorithms that minimize contention by allowing concurrent access to different parts of the data structure.

Unlike most collections, the `size` method of these classes does not necessarily operate in constant time. Determining the current size of one of these collections usually requires traversal.

 NOTE: Some applications use humongous concurrent hash maps, so large that the `size` method is insufficient because it returns an `int`. What is one to do with a map that has over two billion entries? The `mappingCount` method returns the size as a `long`.

The collections return *weakly consistent* iterators. That means that the iterators may or may not reflect all modifications that are made after they were constructed, but they will not return a value twice and they will not throw a `ConcurrentModificationException`.

 NOTE: In contrast, an iterator of a collection in the `java.util` package throws a `ConcurrentModificationException` when the collection has been modified after construction of the iterator.

The concurrent hash map can efficiently support a large number of readers and a fixed number of writers. By default, it is assumed that there are up to 16 *simultaneous* writer threads. There can be many more writer threads, but if more than 16 write at the same time, the others are temporarily blocked. You can specify a higher number in the constructor, but it is unlikely that you will need to.

 NOTE: A hash map keeps all entries with the same hash code in the same "bucket." Some applications use poor hash functions, and as a result all entries end up in a small number of buckets, severely degrading performance. Even generally reasonable hash functions, such as that of the `String` class, can be problematic. For example, an attacker can slow down a program by crafting a large number of strings that hash to the same value. In recent Java versions, the concurrent hash map organizes the buckets as trees, not lists, when the key type implements `Comparable`, guaranteeing $O(\log n)$ performance.

---

java.util.concurrent.ConcurrentLinkedQueue<E> 5

- ConcurrentLinkedQueue<E>()

  constructs an unbounded, nonblocking queue that can be safely accessed by multiple threads.

---

java.util.concurrent.ConcurrentSkipListSet<E> 6

- ConcurrentSkipListSet<E>()
- ConcurrentSkipListSet<E>(Comparator<? super E> comp)

  constructs a sorted set that can be safely accessed by multiple threads. The first constructor requires that the elements implement the Comparable interface.

---

java.util.concurrent.ConcurrentHashMap<K, V> 5
java.util.concurrent.ConcurrentSkipListMap<K, V> 6

- ConcurrentHashMap<K, V>()
- ConcurrentHashMap<K, V>(int initialCapacity)
- ConcurrentHashMap<K, V>(int initialCapacity, float loadFactor, int concurrencyLevel)

  constructs a hash map that can be safely accessed by multiple threads. The default for the initial capacity is 16. If the average load per bucket exceeds the load factor, the table is resized. The default is 0.75. The concurrency level is the estimated number of concurrent writer threads.

- ConcurrentSkipListMap<K, V>()
- ConcurrentSkipListSet<K, V>(Comparator<? super K> comp)

  constructs a sorted map that can be safely accessed by multiple threads. The first constructor requires that the keys implement the Comparable interface.

12.5.3 Atomic Update of Map Entries

The original version of ConcurrentHashMap only had a few methods for atomic updates, which made for somewhat awkward programming. Suppose we want to count how often certain features are observed. As a simple example, suppose multiple threads encounter words, and we want to count their frequencies.

Can we use a ConcurrentHashMap<String, Long>? Consider the code for incrementing a count. Obviously, the following is not thread-safe:

```
Long oldValue = map.get(word);
Long newValue = oldValue == null ? 1 : oldValue + 1;
map.put(word, newValue); // ERROR--might not replace oldValue
```

Another thread might be updating the exact same count at the same time.

> **NOTE:** Some programmers are surprised that a supposedly thread-safe data structure permits operations that are not thread-safe. But there are two entirely different considerations. If multiple threads modify a plain HashMap, they can destroy the internal structure (an array of linked lists). Some of the links may go missing, or even go in circles, rendering the data structure unusable. That will never happen with a ConcurrentHashMap. In the example above, the code for get and put will never corrupt the data structure. But, since the sequence of operations is not atomic, the result is not predictable.

In old versions of Java, it was necessary to use the replace method, which atomically replaces an old value with a new one, provided that no other thread has come before and replaced the old value with something else. You had to keep doing it until the attempt succeeded:

```
do
{
   oldValue = map.get(word);
   newValue = oldValue == null ? 1 : oldValue + 1;
}
while (!map.replace(word, oldValue, newValue));
```

An alternative was to use a ConcurrentHashMap<String, AtomicLong> and the following update code:

```
map.putIfAbsent(word, new AtomicLong());
map.get(word).incrementAndGet();
```

Unfortunately, a new AtomicLong is constructed for each increment, whether or not it is needed.

Nowadays, the Java API provides methods that make atomic updates more convenient. The compute method is called with a key and a function to compute the new value. That function receives the key and the associated value, or null if there is none, and it computes the new value. For example, here is how we can update a map of integer counters:

```
map.compute(word, (k, v) -> v == null ? 1 : v + 1);
```

NOTE: You cannot have `null` values in a `ConcurrentHashMap`. There are many methods that use a `null` value as an indication that a given key is not present in the map.

There are also variants `computeIfPresent` and `computeIfAbsent` that only compute a new value when there is already an old one, or when there isn't yet one. A map of `LongAdder` counters can be updated with

```
map.computeIfAbsent(word, k -> new LongAdder()).increment();
```

That is almost like the call to `putIfAbsent` that you saw before, but the `LongAdder` constructor is only called when a new counter is actually needed.

You often need to do something special when a key is added for the first time. The `merge` method makes this particularly convenient. It has a parameter for the initial value that is used when the key is not yet present. Otherwise, the function that you supplied is called, combining the existing value and the initial value. (Unlike `compute`, the function does *not* process the key.)

```
map.merge(word, 1L, (existingValue, newValue) -> existingValue + newValue);
```

or simply

```
map.merge(word, 1L, Long::sum);
```

It doesn't get more concise than that.

 NOTE: If the function that is passed to `compute` or `merge` returns `null`, the existing entry is removed from the map.

 CAUTION: When you use `compute` or `merge`, keep in mind that the function that you supply should not do a lot of work. While that function runs, some other updates to the map may be blocked. Of course, that function should also not update other parts of the map.

The program in Listing 12.7 uses a concurrent hash map to count all words in the Java files of a directory tree.

Listing 12.7 concurrentHashMap/CHMDemo.java

```
1 package concurrentHashMap;
2
3 import java.io.*;
```

```
4   import java.nio.file.*;
5   import java.util.*;
6   import java.util.concurrent.*;
7   import java.util.stream.*;
8
9   /**
10   * This program demonstrates concurrent hash maps.
11   * @version 1.0 2018-01-04
12   * @author Cay Horstmann
13   */
14  public class CHMDemo
15  {
16     public static ConcurrentHashMap<String, Long> map = new ConcurrentHashMap<>();
17
18     /**
19      * Adds all words in the given file to the concurrent hash map.
20      * @param file a file
21      */
22     public static void process(Path file)
23     {
24        try (var in = new Scanner(file))
25        {
26           while (in.hasNext())
27           {
28              String word = in.next();
29              map.merge(word, 1L, Long::sum);
30           }
31        }
32        catch (IOException e)
33        {
34           e.printStackTrace();
35        }
36     }
37
38     /**
39      * Returns all descendants of a given directory--see Chapters 1 and 2 of Volume II.
40      * @param rootDir the root directory
41      * @return a set of all descendants of the root directory
42      */
43     public static Set<Path> descendants(Path rootDir) throws IOException
44     {
45        try (Stream<Path> entries = Files.walk(rootDir))
46        {
47           return entries.collect(Collectors.toSet());
48        }
49     }
50
```

(Continues)

Listing 12.7 *(Continued)*

```
51   public static void main(String[] args)
52         throws InterruptedException, ExecutionException, IOException
53   {
54      int processors = Runtime.getRuntime().availableProcessors();
55      ExecutorService executor = Executors.newFixedThreadPool(processors);
56      Path pathToRoot = Path.of(".");
57      for (Path p : descendants(pathToRoot))
58      {
59         if (p.getFileName().toString().endsWith(".java"))
60            executor.execute(() -> process(p));
61      }
62      executor.shutdown();
63      executor.awaitTermination(10, TimeUnit.MINUTES);
64      map.forEach((k, v) ->
65         {
66            if (v >= 10)
67               System.out.println(k + " occurs " + v + " times");
68         });
69   }
70 }
```

12.5.4 Bulk Operations on Concurrent Hash Maps

The Java API provides bulk operations on concurrent hash maps that can safely execute even while other threads operate on the map. The bulk operations traverse the map and operate on the elements they find as they go along. No effort is made to freeze a snapshot of the map in time. Unless you happen to know that the map is not being modified while a bulk operation runs, you should treat its result as an approximation of the map's state.

There are three kinds of operations:

- search applies a function to each key and/or value, until the function yields a non-null result. Then the search terminates and the function's result is returned.
- reduce combines all keys and/or values, using a provided accumulation function.
- forEach applies a function to all keys and/or values.

Each operation has four versions:

- *operation*Keys: operates on keys.
- *operation*Values: operates on values.

- *operation*: operates on keys and values.
- *operation*Entries: operates on Map.Entry objects.

With each of the operations, you need to specify a *parallelism threshold*. If the map contains more elements than the threshold, the bulk operation is parallelized. If you want the bulk operation to run in a single thread, use a threshold of Long.MAX_VALUE. If you want the maximum number of threads to be made available for the bulk operation, use a threshold of 1.

Let's look at the search methods first. Here are the versions:

```
U searchKeys(long threshold, BiFunction<? super K, ? extends U> f)
U searchValues(long threshold, BiFunction<? super V, ? extends U> f)
U search(long threshold, BiFunction<? super K, ? super V,? extends U> f)
U searchEntries(long threshold, BiFunction<Map.Entry<K, V>, ? extends U> f)
```

For example, suppose we want to find the first word that occurs more than 1,000 times. We need to search keys and values:

```
String result = map.search(threshold, (k, v) -> v > 1000 ? k : null);
```

Then result is set to the first match, or to null if the search function returns null for all inputs.

The forEach methods have two variants. The first one simply applies a *consumer* function for each map entry, for example

```
map.forEach(threshold,
   (k, v) -> System.out.println(k + " -> " + v));
```

The second variant takes an additional *transformer* function, which is applied first, and its result is passed to the consumer:

```
map.forEach(threshold,
   (k, v) -> k + " -> " + v, // transformer
   System.out::println); // consumer
```

The transformer can be used as a filter. Whenever the transformer returns null, the value is silently skipped. For example, here we only print the entries with large values:

```
map.forEach(threshold,
   (k, v) -> v > 1000 ? k + " -> " + v : null, // filter and transformer
   System.out::println); // the nulls are not passed to the consumer
```

The reduce operations combine their inputs with an accumulation function. For example, here is how you can compute the sum of all values:

```
Long sum = map.reduceValues(threshold, Long::sum);
```

As with forEach, you can also supply a transformer function. Here we compute the length of the longest key:

```
Integer maxlength = map.reduceKeys(threshold,
    String::length, // transformer
    Integer::max); // accumulator
```

The transformer can act as a filter, by returning null to exclude unwanted inputs. Here, we count how many entries have value > 1000:

```
Long count = map.reduceValues(threshold,
    v -> v > 1000 ? 1L : null,
    Long::sum);
```

 NOTE: If the map is empty, or all entries have been filtered out, the reduce operation returns null. If there is only one element, its transformation is returned, and the accumulator is not applied.

There are specializations for int, long, and double outputs with suffixes ToInt, ToLong, and ToDouble. You need to transform the input to a primitive value and specify a default value and an accumulator function. The default value is returned when the map is empty.

```
long sum = map.reduceValuesToLong(threshold,
    Long::longValue, // transformer to primitive type
    0, // default value for empty map
    Long::sum); // primitive type accumulator
```

 CAUTION: These specializations act differently from the object versions where there is only one element to be considered. Instead of returning the transformed element, it is accumulated with the default. Therefore, the default must be the neutral element of the accumulator.

12.5.5 Concurrent Set Views

Suppose you want a large, thread-safe set instead of a map. There is no ConcurrentHashSet class, and you know better than trying to create your own. Of course, you can use a ConcurrentHashMap with bogus values, but then you get a map, not a set, and you can't apply operations of the Set interface.

The static newKeySet method yields a Set<K> that is actually a wrapper around a ConcurrentHashMap<K, Boolean>. (All map values are Boolean.TRUE, but you don't actually care since you just use it as a set.)

```
Set<String> words = ConcurrentHashMap.<String>newKeySet();
```

Of course, if you have an existing map, the keySet method yields the set of keys. That set is mutable. If you remove the set's elements, the keys (and their values) are removed from the map. But it doesn't make sense to add elements to the key set, because there would be no corresponding values to add. There is a second keySet method to ConcurrentHashMap, with a default value, to be used when adding elements to the set:

```
Set<String> words = map.keySet(1L);
words.add("Java");
```

If "Java" wasn't already present in words, it now has a value of one.

12.5.6 Copy on Write Arrays

The CopyOnWriteArrayList and CopyOnWriteArraySet are thread-safe collections in which all mutators make a copy of the underlying array. This arrangement is useful if the threads that iterate over the collection greatly outnumber the threads that mutate it. When you construct an iterator, it contains a reference to the current array. If the array is later mutated, the iterator still has the old array, but the collection's array is replaced. As a consequence, the older iterator has a consistent (but potentially outdated) view that it can access without any synchronization expense.

12.5.7 Parallel Array Algorithms

The Arrays class has a number of parallelized operations. The static Arrays .parallelSort method can sort an array of primitive values or objects. For example,

```
var contents = new String(Files.readAllBytes(
    Path.of("alice.txt")), StandardCharsets.UTF_8); // read file into string
String[] words = contents.split("[\\P{L}]+"); // split along nonletters
Arrays.parallelSort(words);
```

When you sort objects, you can supply a Comparator.

```
Arrays.parallelSort(words, Comparator.comparing(String::length));
```

With all methods, you can supply the bounds of a range, such as

```
values.parallelSort(values.length / 2, values.length); // sort the upper half
```

 NOTE: At first glance, it seems a bit odd that these methods have parallel in their name, since the user shouldn't care how the sorting happens. However, the API designers wanted to make it clear that the sorting is parallelized. That way, users are on notice to avoid comparators with side effects.

The parallelSetAll method fills an array with values that are computed from a function. The function receives the element index and computes the value at that location.

```
Arrays.parallelSetAll(values, i -> i % 10);
   // fills values with 0 1 2 3 4 5 6 7 8 9 0 1 2 . . .
```

Clearly, this operation benefits from being parallelized. There are versions for all primitive type arrays and for object arrays.

Finally, there is a parallelPrefix method that replaces each array element with the accumulation of the prefix for a given associative operation. Huh? Here is an example. Consider the array [1, 2, 3, 4, . . .] and the × operation. After executing Arrays.parallelPrefix(values, (x, y) -> x * y), the array contains

```
[1, 1
    × 2, 1
    × 2
    × 3, 1
    × 2
    × 3
    × 4, . . .]
```

Perhaps surprisingly, this computation can be parallelized. First, join neighboring elements, as indicated here:

```
[1, 1
    × 2, 3, 3
    × 4, 5, 5
    × 6, 7, 7
    × 8]
```

The gray values are left alone. Clearly, one can make this computation in parallel in separate regions of the array. In the next step, update the indicated elements by multiplying them with elements that are one or two positions below:

```
[1, 1 × 2
   , 1 × 2 × 3, 1 × 2 × 3 × 4, 5, 5 × 6, 5 × 6 × 7, 5 × 6 × 7 × 8]
```

This, again, can be done in parallel. After log n steps, the process is complete. This is a win over the straightforward linear computation if sufficient processors are available. On special-purpose hardware, this algorithm is commonly used, and users of such hardware are quite ingenious in adapting it to a variety of problems.

12.5.8 Older Thread-Safe Collections

Ever since the initial release of Java, the Vector and Hashtable classes provided thread-safe implementations of a dynamic array and a hash table. These classes are now considered obsolete, having been replaced by the ArrayList and HashMap classes. Those classes are not thread-safe. Instead, a different mechanism is supplied in the collections library. Any collection class can be made thread-safe by means of a *synchronization wrapper*:

```
List<E> synchArrayList = Collections.synchronizedList(new ArrayList<E>());
Map<K, V> synchHashMap = Collections.synchronizedMap(new HashMap<K, V>());
```

The methods of the resulting collections are protected by a lock, providing thread-safe access.

You should make sure that no thread accesses the data structure through the original unsynchronized methods. The easiest way to ensure this is not to save any reference to the original object. Simply construct a collection and immediately pass it to the wrapper, as we did in our examples.

You still need to use "client-side" locking if you want to *iterate* over the collection while another thread has the opportunity to mutate it:

```
synchronized (synchHashMap)
{
    Iterator<K> iter = synchHashMap.keySet().iterator();
    while (iter.hasNext()) . . .;
}
```

You must use the same code if you use a "for each" loop because the loop uses an iterator. Note that the iterator actually fails with a ConcurrentModificationException if another thread mutates the collection while the iteration is in progress. The synchronization is still required so that the concurrent modification can be reliably detected.

You are usually better off using the collections defined in the java.util.concurrent package instead of the synchronization wrappers. In particular, the ConcurrentHashMap has been carefully implemented so that multiple threads can access it without blocking each other, provided they access different buckets. One exception is an array list that is frequently mutated. In that case, a synchronized ArrayList can outperform a CopyOnWriteArrayList.

---

java.util.Collections 1.2

- static <E> Collection<E> synchronizedCollection(Collection<E> c)
- static <E> List synchronizedList(List<E> c)
- static <E> Set synchronizedSet(Set<E> c)
- static <E> SortedSet synchronizedSortedSet(SortedSet<E> c)
- static <K, V> Map<K, V> synchronizedMap(Map<K, V> c)
- static <K, V> SortedMap<K, V> synchronizedSortedMap(SortedMap<K, V> c)

constructs a view of the collection whose methods are synchronized.

12.6 Tasks and Thread Pools

Constructing a new thread is somewhat expensive because it involves interaction with the operating system. If your program creates a large number of short-lived threads, you should not map each task to a separate thread, but use a *thread pool* instead. A thread pool contains a number of threads that are ready to run. You give a Runnable to the pool, and one of the threads calls the run method. When the run method exits, the thread doesn't die but stays around to serve the next request.

In the following sections, you will see the tools that the Java concurrency framework provides for coordinating concurrent tasks.

12.6.1 Callables and Futures

A Runnable encapsulates a task that runs asynchronously; you can think of it as an asynchronous method with no parameters and no return value. A Callable is similar to a Runnable, but it returns a value. The Callable interface is a parameterized type, with a single method call.

```
public interface Callable<V>
{
   V call() throws Exception;
}
```

The type parameter is the type of the returned value. For example, a Callable<Integer> represents an asynchronous computation that eventually returns an Integer object.

A Future holds the *result* of an asynchronous computation. You start a computation, give someone the Future object, and forget about it. The owner of the Future object can obtain the result when it is ready.

The Future<V> interface has the following methods:

```
V get()
V get(long timeout, TimeUnit unit)
void cancel(boolean mayInterrupt)
boolean isCancelled()
boolean isDone()
```

A call to the first get method blocks until the computation is finished. The second get method also blocks, but it throws a TimeoutException if the call timed out before the computation finished. If the thread running the computation is interrupted, both methods throw an InterruptedException. If the computation has already finished, get returns immediately.

The isDone method returns false if the computation is still in progress, true if it is finished.

You can cancel the computation with the cancel method. If the computation has not yet started, it is canceled and will never start. If the computation is currently in progress, it is interrupted if the mayInterrupt parameter is true.

CAUTION: Canceling a task involves two steps. The underlying thread must be located and interrupted. And the task implementation (in the call method) must sense the interruption and abandon its work. If a Future object does not know on which thread the task is executed, or if the task does not monitor the interrupted status of the thread on which it executes, cancellation will have no effect.

One way to execute a Callable is to use a FutureTask, which implements both the Future and Runnable interfaces, so that you can construct a thread for running it:

```
Callable<Integer> task = . . .;
var futureTask = new FutureTask<Integer>(task);
var t = new Thread(futureTask); // it's a Runnable
t.start();
. . .
Integer result = task.get(); // it's a Future
```

More commonly, you will pass a Callable to an executor. That is the topic of the next section.

---

java.util.concurrent.Callable<V> 5

- V call()

  runs a task that yields a result.

---

java.util.concurrent.Future<V> 5

- V get()
- V get(long time, TimeUnit unit)

  gets the result, blocking until it is available or the given time has elapsed. The second method throws a TimeoutException if it was unsuccessful.

- boolean cancel(boolean mayInterrupt)

  attempts to cancel the execution of this task. If the task has already started and the mayInterrupt parameter is true, it is interrupted. Returns true if the cancellation was successful.

- boolean isCancelled()

  returns true if the task was canceled before it completed.

- boolean isDone()

  returns true if the task completed, through normal completion, cancellation, or an exception.

---

java.util.concurrent.FutureTask<V> 5

- FutureTask(Callable<V> task)
- FutureTask(Runnable task, V result)

  constructs an object that is both a Future<V> and a Runnable.

12.6.2 Executors

The Executors class has a number of static factory methods for constructing thread pools; see Table 12.2 for a summary.

The newCachedThreadPool method constructs a thread pool that executes each task immediately, using an existing idle thread when available and creating a new thread otherwise. The newFixedThreadPool method constructs a thread pool with a fixed size. If more tasks are submitted than there are idle threads, the un-served tasks are placed on a queue. They are run when other tasks have completed. The newSingleThreadExecutor is a degenerate pool of size 1 where a single thread executes the submitted tasks, one after another. These three methods return an object of the ThreadPoolExecutor class that implements the ExecutorService interface.

Use a cached thread pool when you have threads that are short-lived or spend a lot of time blocking. However, if you have threads that are working hard without blocking, you don't want to run a large number of them together.

Table 12.2 Executors Factory Methods

Method	Description
newCachedThreadPool	New threads are created as needed; idle threads are kept for 60 seconds.
newFixedThreadPool	The pool contains a fixed set of threads; idle threads are kept indefinitely.
newWorkStealingPool	A pool suitable for "fork-join" tasks (see Section 12.6.4) in which complex tasks are broken up into simpler tasks and idle threads "steal" simpler tasks.
newSingleThreadExecutor	A "pool" with a single thread that executes the submitted tasks sequentially.
newScheduledThreadPool	A fixed-thread pool for scheduled execution.
newSingleThreadScheduledExecutor	A single-thread "pool" for scheduled execution.

For optimum speed, the number of concurrent threads is the number of processor cores. In such a situation, you should use a fixed thread pool that bounds the total number of concurrent threads.

The single-thread executor is useful for performance analysis. If you temporarily replace a cached or fixed thread pool with a single-thread pool, you can measure how much slower your application runs without the benefit of concurrency.

NOTE: Java EE provides a ManagedExecutorService subclass that is suitable for concurrent tasks in a Java EE environment. Similarly, web frameworks such as Play provide executor services that are intended for tasks within the framework.

You can submit a Runnable or Callable to an ExecutorService with one of the following methods:

```
Future<T> submit(Callable<T> task)
Future<?> submit(Runnable task)
Future<T> submit(Runnable task, T result)
```

The pool will run the submitted task at its earliest convenience. When you call submit, you get back a Future object that you can use to get the result or cancel the task.

The second submit method returns an odd-looking Future<?>. You can use such an object to call isDone, cancel, or isCancelled, but the get method simply returns null upon completion.

The third version of submit yields a Future whose get method returns the given result object upon completion.

When you are done with a thread pool, call shutdown. This method initiates the shutdown sequence for the pool. An executor that is shut down accepts no new tasks. When all tasks are finished, the threads in the pool die. Alternatively, you can call shutdownNow. The pool then cancels all tasks that have not yet begun.

Here, in summary, is what you do to use a thread pool:

1. Call the static newCachedThreadPool or newFixedThreadPool method of the Executors class.
2. Call submit to submit Callable or Runnable objects.
3. Hang on to the returned Future objects so that you can get the results or cancel the tasks.
4. Call shutdown when you no longer want to submit any tasks.

The ScheduledExecutorService interface has methods for scheduled or repeated execution of tasks. It is a generalization of java.util.Timer that allows for thread pooling. The newScheduledThreadPool and newSingleThreadScheduledExecutor methods of the Executors class return objects that implement the ScheduledExecutorService interface.

You can schedule a Runnable or Callable to run once, after an initial delay. You can also schedule a Runnable to run periodically. See the API notes for details.

---

java.util.concurrent.Executors 5

- ExecutorService newCachedThreadPool()

  returns a cached thread pool that creates threads as needed and terminates threads that have been idle for 60 seconds.
- ExecutorService newFixedThreadPool(int threads)

  returns a thread pool that uses the given number of threads to execute tasks.
- ExecutorService newSingleThreadExecutor()

  returns an executor that executes tasks sequentially in a single thread.

(Continues)

---

java.util.concurrent.Executors 5 *(Continued)*

- `ScheduledExecutorService newScheduledThreadPool(int threads)`

  returns a thread pool that uses the given number of threads to schedule tasks.

- `ScheduledExecutorService newSingleThreadScheduledExecutor()`

  returns an executor that schedules tasks in a single thread.

---

java.util.concurrent.ExecutorService 5

- `Future<T> submit(Callable<T> task)`
- `Future<T> submit(Runnable task, T result)`
- `Future<?> submit(Runnable task)`

  submits the given task for execution.

- `void shutdown()`

  shuts down the service, completing the already submitted tasks but not accepting new submissions.

---

java.util.concurrent.ThreadPoolExecutor 5

- `int getLargestPoolSize()`

  returns the largest size of the thread pool during the life of this executor.

---

java.util.concurrent.ScheduledExecutorService 5

- `ScheduledFuture<V> schedule(Callable<V> task, long time, TimeUnit unit)`
- `ScheduledFuture<?> schedule(Runnable task, long time, TimeUnit unit)`

  schedules the given task after the given time has elapsed.

- `ScheduledFuture<?> scheduleAtFixedRate(Runnable task, long initialDelay, long period, TimeUnit unit)`

  schedules the given task to run periodically, every period units, after the initial delay has elapsed.

- `ScheduledFuture<?> scheduleWithFixedDelay(Runnable task, long initialDelay, long delay, TimeUnit unit)`

  schedules the given task to run periodically, with delay units between completion of one invocation and the start of the next, after the initial delay has elapsed.

12.6.3 Controlling Groups of Tasks

You have seen how to use an executor service as a thread pool to increase the efficiency of task execution. Sometimes, an executor is used for a more tactical reason—simply to control a group of related tasks. For example, you can cancel all tasks in an executor with the shutdownNow method.

The invokeAny method submits all objects in a collection of Callable objects and returns the result of a completed task. You don't know which task that is—presumably, it is the one that finished most quickly. Use this method for a search problem in which you are willing to accept any solution. For example, suppose that you need to factor a large integer—a computation that is required for breaking the RSA cipher. You could submit a number of tasks, each attempting a factorization with numbers in a different range. As soon as one of these tasks has an answer, your computation can stop.

The invokeAll method submits all objects in a collection of Callable objects, blocks until all of them complete, and returns a list of Future objects that represent the solutions to all tasks. You can process the results of the computation, when they are available, like this:

```
List<Callable<T>> tasks = . . .;
List<Future<T>> results = executor.invokeAll(tasks);
for (Future<T> result : results)
   processFurther(result.get());
```

In the for loop, the first call result.get() blocks until the first result is available. That is not a problem if all tasks finish in about the same time. However, it may be worth obtaining the results in the order in which they are available. This can be arranged with the ExecutorCompletionService.

Start with an executor, obtained in the usual way. Then construct an ExecutorCompletionService. Submit tasks to the completion service. The service manages a blocking queue of Future objects, containing the results of the submitted tasks as they become available. Thus, a more efficient organization for the preceding computation is the following:

```
var service = new ExecutorCompletionService<T>(executor);
for (Callable<T> task : tasks) service.submit(task);
for (int i = 0; i < tasks.size(); i++)
   processFurther(service.take().get());
```

The program in Listing 12.8 shows how to use callables and executors. In the first computation, we count how many files in a directory tree contain a given word. We make a separate task for each file:

```
Set<Path> files = descendants(Path.of(start));
var tasks = new ArrayList<Callable<Long>>();
```

```
for (Path file : files)
{
   Callable<Long> task = () -> occurrences(word, file);
   tasks.add(task);
}
```

Then we pass the tasks to an executor service:

```
ExecutorService executor = Executors.newCachedThreadPool();
List<Future<Long>> results = executor.invokeAll(tasks);
```

To get the combined count, we add all results, blocking until they are available:

```
long total = 0;
for (Future<Long> result : results)
   total += result.get();
```

The program also displays the time spent during the search. Unzip the source code for the JDK somewhere and run the search. Then replace the executor service with a single-thread executor and try again to see whether the concurrent computation was faster.

In the second part of the program, we search for the first file that contains the given word. We use invokeAny to parallelize the search. Here, we have to be more careful about formulating the tasks. The invokeAny method terminates as soon as any task *returns*. So we cannot have the search tasks return a boolean to indicate success or failure. We don't want to stop searching when a task failed. Instead, a failing task throws a NoSuchElementException. Also, when one task has succeeded, the others are canceled. Therefore, we monitor the interrupted status. If the underlying thread is interrupted, the search task prints a message before terminating, so that you can see that the cancellation is effective.

```
public static Callable<Path> searchForTask(String word, Path path)
{
   return () -> {
      try (var in = new Scanner(path))
      {
         while (in.hasNext())
         {
            if (in.next().equals(word)) return path;
            if (Thread.currentThread().isInterrupted())
            {
               System.out.println("Search in " + path + " canceled.");
               return null;
            }
         }
      }
```

```
            throw new NoSuchElementException();
         }
      };
   }
```

For informational purposes, this program prints out the largest pool size during execution. This information is not available through the ExecutorService interface. For that reason, we had to cast the pool object to the ThreadPoolExecutor class.

 TIP: As you read through this program, you can appreciate how useful executor services are. In your own programs, you should use executor services to manage threads instead of launching threads individually.

Listing 12.8 executors/ExecutorDemo.java

```
1  package executors;
2
3  import java.io.*;
4  import java.nio.file.*;
5  import java.time.*;
6  import java.util.*;
7  import java.util.concurrent.*;
8  import java.util.stream.*;
9
10 /**
11  * This program demonstrates the Callable interface and executors.
12  * @version 1.0 2018-01-04
13  * @author Cay Horstmann
14  */
15 public class ExecutorDemo
16 {
17    /**
18     * Counts occurrences of a given word in a file.
19     * @return the number of times the word occurs in the given word
20     */
21    public static long occurrences(String word, Path path)
22    {
23       try (var in = new Scanner(path))
24       {
25          int count = 0;
26          while (in.hasNext())
27             if (in.next().equals(word)) count++;
28          return count;
29       }
```

```
30      catch (IOException ex)
31      {
32         return 0;
33      }
34   }
35
36   /**
37    * Returns all descendants of a given directory--see Chapters 1 and 2 of Volume II.
38    * @param rootDir the root directory
39    * @return a set of all descendants of the root directory
40    */
41   public static Set<Path> descendants(Path rootDir) throws IOException
42   {
43      try (Stream<Path> entries = Files.walk(rootDir))
44      {
45         return entries.filter(Files::isRegularFile)
46            .collect(Collectors.toSet());
47      }
48   }
49
50   /**
51    * Yields a task that searches for a word in a file.
52    * @param word the word to search
53    * @param path the file in which to search
54    * @return the search task that yields the path upon success
55    */
56   public static Callable<Path> searchForTask(String word, Path path)
57   {
58      return () -> {
59         try (var in = new Scanner(path))
60         {
61            while (in.hasNext())
62            {
63               if (in.next().equals(word)) return path;
64               if (Thread.currentThread().isInterrupted())
65               {
66                  System.out.println("Search in " + path + " canceled.");
67                  return null;
68               }
69            }
70            throw new NoSuchElementException();
71         }
72      };
73   }
74
```

(Continues)

Listing 12.8 *(Continued)*

```java
75    public static void main(String[] args)
76          throws InterruptedException, ExecutionException, IOException
77    {
78       try (var in = new Scanner(System.in))
79       {
80          System.out.print("Enter base directory (e.g. /opt/jdk-9-src): ");
81          String start = in.nextLine();
82          System.out.print("Enter keyword (e.g. volatile): ");
83          String word = in.nextLine();
84
85          Set<Path> files = descendants(Path.of(start));
86          var tasks = new ArrayList<Callable<Long>>();
87          for (Path file : files)
88          {
89             Callable<Long> task = () -> occurrences(word, file);
90             tasks.add(task);
91          }
92          ExecutorService executor = Executors.newCachedThreadPool();
93          // use a single thread executor instead to see if multiple threads
94          // speed up the search
95          // ExecutorService executor = Executors.newSingleThreadExecutor();
96
97          Instant startTime = Instant.now();
98          List<Future<Long>> results = executor.invokeAll(tasks);
99          long total = 0;
100         for (Future<Long> result : results)
101            total += result.get();
102         Instant endTime = Instant.now();
103         System.out.println("Occurrences of " + word + ": " + total);
104         System.out.println("Time elapsed: "
105            + Duration.between(startTime, endTime).toMillis() + " ms");
106
107         var searchTasks = new ArrayList<Callable<Path>>();
108         for (Path file : files)
109            searchTasks.add(searchForTask(word, file));
110         Path found = executor.invokeAny(searchTasks);
111         System.out.println(word + " occurs in: " + found);
112
113         if (executor instanceof ThreadPoolExecutor) // the single thread executor isn't
114            System.out.println("Largest pool size: "
115               + ((ThreadPoolExecutor) executor).getLargestPoolSize());
116         executor.shutdown();
117      }
118   }
119 }
```

---

java.util.concurrent.ExecutorService 5

- T invokeAny(Collection<Callable<T>> tasks)
- T invokeAny(Collection<Callable<T>> tasks, long timeout, TimeUnit unit)

  executes the given tasks and returns the result of one of them. The second method throws a TimeoutException if a timeout occurs.

- List<Future<T>> invokeAll(Collection<Callable<T>> tasks)
- List<Future<T>> invokeAll(Collection<Callable<T>> tasks, long timeout, TimeUnit unit)

  executes the given tasks and returns the results of all of them. The second method throws a TimeoutException if a timeout occurs.

---

java.util.concurrent.ExecutorCompletionService<V> 5

- ExecutorCompletionService(Executor e)

  constructs an executor completion service that collects the results of the given executor.

- Future<V> submit(Callable<V> task)
- Future<V> submit(Runnable task, V result)

  submits a task to the underlying executor.

- Future<V> take()

  removes the next completed result, blocking if no completed results are available.

- Future<V> poll()
- Future<V> poll(long time, TimeUnit unit)

  removes and returns the next completed result, or returns null if no completed results are available. The second method waits for the given time.

12.6.4 The Fork–Join Framework

Some applications use a large number of threads that are mostly idle. An example would be a web server that uses one thread per connection. Other applications use one thread per processor core, in order to carry out computationally intensive tasks, such as image or video processing. The fork-join framework, which appeared in Java 7, is designed to support the latter. Suppose you have a processing task that naturally decomposes into subtasks, like this:

```
if (problemSize < threshold)
    solve problem directly
else
{
```

```
        break problem into subproblems
        recursively solve each subproblem
        combine the results
    }
```

One example is image processing. To enhance an image, you can transform the top half and the bottom half. If you have enough idle processors, those operations can run in parallel. (You will need to do a bit of extra work along the strip that separates the two halves, but that's a technical detail.)

Here, we discuss a simpler example. Suppose we want to count how many elements of an array fulfill a particular property. We cut the array in half, compute the counts of each half, and add them up.

To put the recursive computation in a form that is usable by the framework, supply a class that extends RecursiveTask<T> (if the computation produces a result of type T) or RecursiveAction (if it doesn't produce a result). Override the compute method to generate and invoke subtasks, and to combine their results.

```
class Counter extends RecursiveTask<Integer>
{
    . . .
    protected Integer compute()
    {
        if (to - from < THRESHOLD)
        {
            solve problem directly
        }
        else
        {
            int mid = (from + to) / 2;
            var first = new Counter(values, from, mid, filter);
            var second = new Counter(values, mid, to, filter);
            invokeAll(first, second);
            return first.join() + second.join();
        }
    }
}
```

Here, the invokeAll method receives a number of tasks and blocks until all of them have completed. The join method yields the result. Here, we apply join to each subtask and return the sum.

 NOTE: There is also a get method for getting the current result, but it is less attractive since it can throw checked exceptions that we are not allowed to throw in the compute method.

Listing 12.9 shows the complete example.

Behind the scenes, the fork-join framework uses an effective heuristic, called *work stealing*, for balancing the workload among available threads. Each worker thread has a deque (double-ended queue) for tasks. A worker thread pushes subtasks onto the head of its own deque. (Only one thread accesses the head, so no locking is required.) When a worker thread is idle, it "steals" a task from the tail of another deque. Since large subtasks are at the tail, such stealing is rare.

 CAUTION: Fork-join pools are optimized for nonblocking workloads. If you add many blocking tasks into a fork-join pool, you can starve it. It is possible to overcome this by having tasks implement the ForkJoinPool.ManagedBlocker interface, but this is an advanced technique that we won't discuss.

Listing 12.9 forkJoin/ForkJoinTest.java

```
1  package forkJoin;
2
3  import java.util.concurrent.*;
4  import java.util.function.*;
5
6  /**
7   * This program demonstrates the fork-join framework.
8   * @version 1.01 2015-06-21
9   * @author Cay Horstmann
10  */
11 public class ForkJoinTest
12 {
13    public static void main(String[] args)
14    {
15       final int SIZE = 10000000;
16       var numbers = new double[SIZE];
17       for (int i = 0; i < SIZE; i++) numbers[i] = Math.random();
18       var counter = new Counter(numbers, 0, numbers.length, x -> x > 0.5);
19       var pool = new ForkJoinPool();
20       pool.invoke(counter);
21       System.out.println(counter.join());
22    }
23 }
24
25 class Counter extends RecursiveTask<Integer>
26 {
```

(Continues)

Listing 12.9 *(Continued)*

```
27    public static final int THRESHOLD = 1000;
28    private double[] values;
29    private int from;
30    private int to;
31    private DoublePredicate filter;
32
33    public Counter(double[] values, int from, int to, DoublePredicate filter)
34    {
35       this.values = values;
36       this.from = from;
37       this.to = to;
38       this.filter = filter;
39    }
40
41    protected Integer compute()
42    {
43       if (to - from < THRESHOLD)
44       {
45          int count = 0;
46          for (int i = from; i < to; i++)
47          {
48             if (filter.test(values[i])) count++;
49          }
50          return count;
51       }
52       else
53       {
54          int mid = (from + to) / 2;
55          var first = new Counter(values, from, mid, filter);
56          var second = new Counter(values, mid, to, filter);
57          invokeAll(first, second);
58          return first.join() + second.join();
59       }
60    }
61 }
```

12.7 Asynchronous Computations

So far, our approach to concurrent computation has been to break up a task, and then wait until all pieces have completed. But waiting is not always a good idea. In the following sections, you will see how to implement wait-free, or *asynchronous*, computations.

12.7.1 Completable Futures

When you have a Future object, you need to call get to obtain the value, blocking until the value is available. The CompletableFuture class implements the Future interface, and it provides a second mechanism for obtaining the result. You register a *callback* that will be invoked (in some thread) with the result once it is available.

```
CompletableFuture<String> f = . . .;
f.thenAccept(s -> Process the result string s);
```

In this way, you can process the result without blocking once it is available.

There are a few API methods that return CompletableFuture objects. For example, you can fetch a web page asynchronously with the experimental HttpClient class that you will encounter in Chapter 4 of Volume II:

```
HttpClient client = HttpClient.newHttpClient();
HttpRequest request = HttpRequest.newBuilder(URI.create(urlString)).GET().build();
CompletableFuture<HttpResponse<String>> f = client.sendAsync(
    request, BodyHandler.asString());
```

It is nice if there is a method that produces a ready-made CompletableFuture, but most of the time, you need to make your own. To run a task asynchronously and obtain a CompletableFuture, you don't submit it directly to an executor service. Instead, you call the static method CompletableFuture.supplyAsync. Here is how to read the web page without the benefit of the HttpClient class:

```
public CompletableFuture<String> readPage(URL url)
{
    return CompletableFuture.supplyAsync(() ->
        {
            try
            {
                return new String(url.openStream().readAllBytes(), "UTF-8");
            }
            catch (IOException e)
            {
                throw new UncheckedIOException(e);
            }
        }, executor);
}
```

If you omit the executor, the task is run on a default executor (namely the executor returned by ForkJoinPool.commonPool()). You usually don't want to do that.

CAUTION: Note that the first argument of the supplyAsync method is a Supplier<T>, not a Callable<T>. Both interfaces describe functions with no arguments and a return value of type T, but a Supplier function cannot throw a checked exception. As you can see from the code above, that was not an inspired choice.

A CompletableFuture can complete in two ways: either with a result, or with an uncaught exception. In order to handle both cases, use the whenComplete method. The supplied function is called with the result (or null if none) and the exception (or null if none).

```
f.whenComplete((s, t) -> {
   if (t == null) { Process the result s; }
   else { Process the Throwable t; }
});
```

The CompletableFuture is called completable because you can manually set a completion value. (In other concurrency libraries, such an object is called a *promise*.) Of course, when you create a CompletableFuture with supplyAsync, the completion value is implicitly set when the task has finished. But setting the result explicitly gives you additional flexibility. For example, two tasks can work simultaneously on computing an answer:

```
var f = new CompletableFuture<Integer>();
executor.execute(() ->
   {
       int n = workHard(arg);
       f.complete(n);
   });
executor.execute(() ->
   {
       int n = workSmart(arg);
       f.complete(n);
   });
```

To instead complete a future with an exception, call

```
Throwable t = . . .;
f.completeExceptionally(t);
```

 NOTE: It is safe to call complete or completeExceptionally on the same future in multiple threads. If the future is already completed, these calls have no effect.

The isDone method tells you whether a Future object has been completed (normally or with an exception). In the preceding example, the workHard and workSmart methods can use that information to stop working when the result has been determined by the other method.

 CAUTION: Unlike a plain Future, the computation of a CompletableFuture is not interrupted when you invoke its cancel method. Canceling simply sets the Future object to be completed exceptionally, with a CancellationException. In general, this makes sense since a CompletableFuture may not have a single thread that is responsible for its completion. However, this restriction also applies to CompletableFuture instances returned by methods such as supplyAsync, which could in principle be interrupted.

12.7.2 Composing Completable Futures

Nonblocking calls are implemented through callbacks. The programmer registers a callback for the action that should occur after a task completes. Of course, if the next action is also asynchronous, the next action after that is in a different callback. Even though the programmer thinks in terms of "first do step 1, then step 2, then step 3," the program logic can become dispersed in "callback hell." It gets even worse when one has to add error handling. Suppose step 2 is "the user logs in." You may need to repeat that step since the user can mistype the credentials. Trying to implement such a control flow in a set of callbacks, or to understand it once it has been implemented, can be quite challenging.

The CompletableFuture class solves this problem by providing a mechanism for *composing* asynchronous tasks into a processing pipeline.

For example, suppose we want to extract all images from a web page. Let's say we have a method

```
public void CompletableFuture<String> readPage(URL url)
```

that yields the text of a web page when it becomes available. If the method

```
public List<URL> getImageURLs(String page)
```

yields the URLs of images in an HTML page, you can schedule it to be called when the page is available:

```
CompletableFuture<String> contents = readPage(url);
CompletableFuture<List<URL>> imageURLs = contents.thenApply(this::getLinks);
```

The thenApply method doesn't block either. It returns another future. When the first future has completed, its result is fed to the getImageURLs method, and the return value of that method becomes the final result.

With completable futures, you just specify what you want to have done and in which order. It won't all happen right away, of course, but what is important is that all the code is in one place.

Conceptually, `CompletableFuture` is a simple API, but there are many variants of methods for composing completable futures. Let us first look at those that deal with a single future (see Table 12.3). (For each method shown, there are also two `Async` variants that I don't show. One of them uses a shared `ForkJoinPool`, and the other has an `Executor` parameter.) In the table, I use a shorthand notation for the ponderous functional interfaces, writing `T -> U` instead of `Function<? super T, U>`. These aren't actual Java types, of course.

You have already seen the `thenApply` method. Suppose `f` is a function that receives values of type `T` and returns values of type `U`. The calls

```
CompletableFuture<U> future.thenApply(f);
CompletableFuture<U> future.thenApplyAsync(f);
```

return a future that applies the function `f` to the result of `future` when it is available. The second call runs `f` in yet another thread.

The `thenCompose` method, instead of taking a function mapping the type `T` to the type `U`, receives a function mapping `T` to `CompletableFuture<U>`. That sounds rather abstract, but it can be quite natural. Consider the action of reading a web page from a given URL. Instead of supplying a method

```
public String blockingReadPage(URL url)
```

it is more elegant to have that method return a future:

```
public CompletableFuture<String> readPage(URL url)
```

Now, suppose we have another method that gets the URL from user input, perhaps from a dialog that won't reveal the answer until the user has clicked the OK button. That, too, is an event in the future:

```
public CompletableFuture<URL> getURLInput(String prompt)
```

Here we have two functions `T -> CompletableFuture<U>` and `U -> CompletableFuture<V>`. Clearly, they compose to a function `T -> CompletableFuture<V>` if the second function is called when the first one has completed. That is exactly what `thenCompose` does.

In the preceding section, you saw the `whenComplete` method for handling exceptions. There is also a `handle` method that requires a function processing the result or exception and computing a new result. In many cases, it is simpler to call the `exceptionally` method instead. That method computes a dummy value when an exception occurs:

```
CompletableFuture<List<URL>> imageURLs = readPage(url)
    .exceptionally(ex -> "<html></html>")
    .thenApply(this::getImageURLs)
```

You can handle a timeout in the same way:

```
CompletableFuture<List<URL>> imageURLs = readPage(url)
    .completeOnTimeout("<html></html>", 30, TimeUnit.SECONDS)
    .thenApply(this::getImageURLs)
```

Alternatively, you can throw an exception on timeout:

```
CompletableFuture<String> = readPage(url).orTimeout(30, TimeUnit.SECONDS)
```

The methods in Table 12.3 with void result are normally used at the end of a processing pipeline.

Table 12.3 Adding an Action to a CompletableFuture<T> Object

Method	Parameter	Description
thenApply	T -> U	Apply a function to the result.
thenAccept	T -> void	Like thenApply, but with void result.
thenCompose	T -> CompletableFuture<U>	Invoke the function on the result and execute the returned future.
handle	(T, Throwable) -> U	Process the result or error and yield a new result.
whenComplete	(T, Throwable) -> void	Like handle, but with void result.
exceptionally	Throwable -> T	Compute a result from the error.
completeOnTimeout	T, long, TimeUnit	Yield the given value as the result in case of timeout.
orTimeout	long, TimeUnit	Yield a TimeoutException in case of timeout.
thenRun	Runnable	Execute the Runnable with void result.

Now let us turn to methods that combine multiple futures (see Table 12.4).

The first three methods run a CompletableFuture<T> and a CompletableFuture<U> action concurrently and combine the results.

The next three methods run two CompletableFuture<T> actions concurrently. As soon as one of them finishes, its result is passed on, and the other result is ignored.

Finally, the static allOf and anyOf methods take a variable number of completable futures and yield a CompletableFuture<Void> that completes when all of them, or any one of them, completes. The allOf method does not yield a result. The anyOf method does *not* terminate the remaining tasks.

Table 12.4 Combining Multiple Composition Objects

Method	Parameters	Description
thenCombine	CompletableFuture<U>, (T, U) -> V	Execute both and combine the results with the given function.
thenAcceptBoth	CompletableFuture<U>, (T, U) -> void	Like thenCombine, but with void result.
runAfterBoth	CompletableFuture<?>, Runnable	Execute the runnable after both complete.
applyToEither	CompletableFuture<T>, T -> V	When a result is available from one or the other, pass it to the given function.
acceptEither	CompletableFuture<T>, T -> void	Like applyToEither, but with void result.
runAfterEither	CompletableFuture<?>, Runnable	Execute the runnable after one or the other completes.
static allOf	CompletableFuture<?>...	Complete with void result after all given futures complete.
static anyOf	CompletableFuture<?>...	Complete with void result after any of the given futures completes.

 NOTE: Technically speaking, the methods in this section accept parameters of type CompletionStage, not CompletableFuture. The CompletionStage interface describes how to compose asynchronous computations, whereas the Future interface focuses on the result of a computation. A CompletableFuture is both a CompletionStage and a Future.

Listing 12.10 shows a complete program that reads a web page, scans it for images, loads the images and saves them locally. Note how all time-consuming methods return a CompletableFuture. To kick off the asynchronous computation, we use a little trick. Rather than calling the readPage method directly, we make a completed future with the URL argument, and then compose that future with this::readPage. That way, the pipeline has a very uniform appearance:

```
CompletableFuture.completedFuture(url)
   .thenComposeAsync(this::readPage, executor)
   .thenApply(this::getImageURLs)
   .thenCompose(this::getImages)
   .thenAccept(this::saveImages);
```

Listing 12.10 completableFutures/CompletableFutureDemo.java

```
1   package completableFutures;
2
3   import java.awt.image.*;
4   import java.io.*;
5   import java.net.*;
6   import java.nio.charset.*;
7   import java.util.*;
8   import java.util.concurrent.*;
9   import java.util.regex.*;
10
11  import javax.imageio.*;
12
13  public class CompletableFutureDemo
14  {
15     private static final Pattern IMG_PATTERN = Pattern.compile(
16        "[<]\\s*[iI][mM][gG]\\s*[^>]*[sS][rR][cC]\\s*[=]\\s*['\"]([^'\"]*)['\"][^>]*[>]");
17     private ExecutorService executor = Executors.newCachedThreadPool();
18     private URL urlToProcess;
19
20     public CompletableFuture<String> readPage(URL url)
21     {
22        return CompletableFuture.supplyAsync(() ->
23           {
24              try
25              {
26                 var contents = new String(url.openStream().readAllBytes(),
27                    StandardCharsets.UTF_8);
28                 System.out.println("Read page from " + url);
29                 return contents;
30              }
31              catch (IOException e)
32              {
33                 throw new UncheckedIOException(e);
34              }
35           }, executor);
36     }
37
38     public List<URL> getImageURLs(String webpage) // not time-consuming
39     {
```

(Continues)

Listing 12.10 *(Continued)*

```
40        try
41        {
42           var result = new ArrayList<URL>();
43           Matcher matcher = IMG_PATTERN.matcher(webpage);
44           while (matcher.find())
45           {
46              var url = new URL(urlToProcess, matcher.group(1));
47              result.add(url);
48           }
49           System.out.println("Found URLs: " + result);
50           return result;
51        }
52        catch (IOException e)
53        {
54           throw new UncheckedIOException(e);
55        }
56     }
57
58     public CompletableFuture<List<BufferedImage>> getImages(List<URL> urls)
59     {
60        return CompletableFuture.supplyAsync(() ->
61        {
62           try
63           {
64              var result = new ArrayList<BufferedImage>();
65              for (URL url : urls)
66              {
67                 result.add(ImageIO.read(url));
68                 System.out.println("Loaded " + url);
69              }
70              return result;
71           }
72           catch (IOException e)
73           {
74              throw new UncheckedIOException(e);
75           }
76        }, executor);
77     }
78
79     public void saveImages(List<BufferedImage> images)
80     {
81        System.out.println("Saving " + images.size() + " images");
82        try
83        {
84           for (int i = 0; i < images.size(); i++)
85           {
```

```
86          String filename = "/tmp/image" + (i + 1) + ".png";
87          ImageIO.write(images.get(i), "PNG", new File(filename));
88       }
89     }
90     catch (IOException e)
91     {
92        throw new UncheckedIOException(e);
93     }
94     executor.shutdown();
95  }
96
97  public void run(URL url)
98        throws IOException, InterruptedException
99  {
100    urlToProcess = url;
101    CompletableFuture.completedFuture(url)
102       .thenComposeAsync(this::readPage, executor)
103       .thenApply(this::getImageURLs)
104       .thenCompose(this::getImages)
105       .thenAccept(this::saveImages);
106
107    /*
108    // or use the experimental HTTP client:
109
110    HttpClient client = HttpClient.newBuilder().executor(executor).build();
111    HttpRequest request = HttpRequest.newBuilder(urlToProcess.toURI()).GET()
112       .build();
113    client.sendAsync(request, BodyProcessor.asString())
114       .thenApply(HttpResponse::body).thenApply(this::getImageURLs)
115       .thenCompose(this::getImages).thenAccept(this::saveImages);
116    */
117  }
118
119  public static void main(String[] args)
120        throws IOException, InterruptedException
121  {
122    new CompletableFutureDemo().run(new URL("http://horstmann.com/index.html"));
123  }
124 }
```

12.7.3 Long-Running Tasks in User Interface Callbacks

One of the reasons to use threads is to make your programs more responsive.
This is particularly important in an application with a user interface. When
your program needs to do something time-consuming, you cannot do the
work in the user-interface thread, or the user interface will be frozen. Instead,
fire up another worker thread.

For example, if you want to read a file when the user clicks a button, don't do this:

```
var open = new JButton("Open");
open.addActionListener(event ->
   { // BAD--long-running action is executed on UI thread
      var in = new Scanner(file);
      while (in.hasNextLine())
      {
         String line = in.nextLine();
         . . .
      }
   });
```

Instead, do the work in a separate thread.

```
open.addActionListener(event ->
   { // GOOD--long-running action in separate thread
      Runnable task = () ->
         {
            var in = new Scanner(file);
            while (in.hasNextLine())
            {
               String line = in.nextLine();
               . . .
            }
         };
      executor.execute(task);
   });
```

However, you cannot directly update the user interface from the worker thread that executes the long-running task. User interfaces such as Swing, JavaFX, or Android are not thread-safe. You cannot manipulate user interface elements from multiple threads, or they risk becoming corrupted. In fact, JavaFX and Android check for this, and throw an exception if you try to access the user interface from a thread other than the UI thread.

Therefore, you need to schedule any UI updates to happen on the UI thread. Each user interface library provides some mechanism to schedule a Runnable for execution on the UI thread. For example, in Swing, you call

```
EventQueue.invokeLater(() -> label.setText(percentage + "% complete"));
```

It is tedious to implement user feedback in a worker thread, so each user interface library provides some kind of helper class for managing the details, such as SwingWorker in Swing, Task in JavaFX, and AsyncTask in Android. You specify actions for the long-running task (which is run on a separate thread), as well as progress updates and the final disposition (which are run on the UI thread).

The program in Listing 12.11 has commands for loading a text file and for canceling the file loading process. You should try the program with a long file, such as the full text of *The Count of Monte Cristo*, supplied in the gutenberg directory of the book's companion code. The file is loaded in a separate thread. While the file is being read, the Open menu item is disabled and the Cancel item is enabled (see Figure 12.6). After each line is read, a line counter in the status bar is updated. After the reading process is complete, the Open menu item is reenabled, the Cancel item is disabled, and the status line text is set to Done.

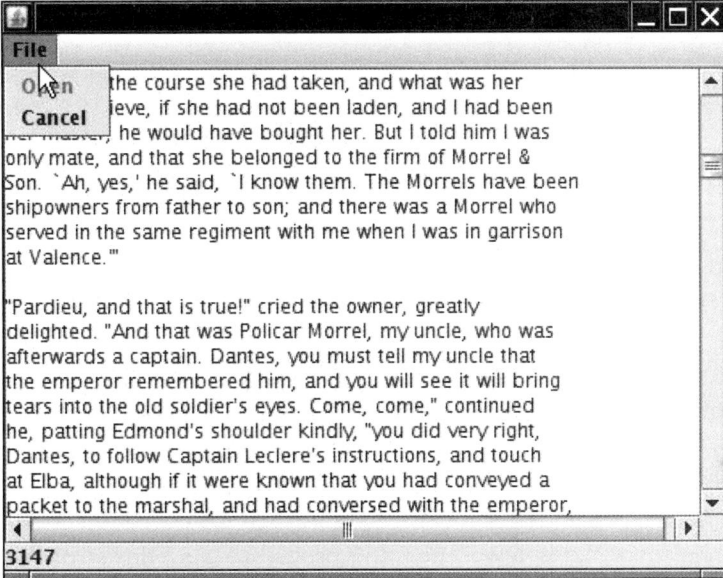

Figure 12.6 Loading a file in a separate thread

This example shows the typical UI activities of a background task:

• After each work unit, update the UI to show progress.
• After the work is finished, make a final change to the UI.

The SwingWorker class makes it easy to implement such a task. Override the doInBackground method to do the time-consuming work and occasionally call publish to communicate work progress. This method is executed in a worker thread. The publish method causes a process method to execute in the event dispatch thread to deal with the progress data. When the work is complete, the done method is called in the event dispatch thread so that you can finish updating the UI.

Whenever you want to do some work in the worker thread, construct a new worker. (Each worker object is meant to be used only once.) Then call the execute method. You will typically call execute on the event dispatch thread, but that is not a requirement.

It is assumed that a worker produces a result of some kind; therefore, SwingWorker<T, V> implements Future<T>. This result can be obtained by the get method of the Future interface. Since the get method blocks until the result is available, you don't want to call it immediately after calling execute. It is a good idea to call it only when you know that the work has been completed. Typically, you call get from the done method. (There is no requirement to call get. Sometimes, processing the progress data is all you need.)

Both the intermediate progress data and the final result can have arbitrary types. The SwingWorker class has these types as type parameters. A SwingWorker<T, V> produces a result of type T and progress data of type V.

To cancel the work in progress, use the cancel method of the Future interface. When the work is canceled, the get method throws a CancellationException.

As already mentioned, the worker thread's call to publish will cause calls to process on the event dispatch thread. For efficiency, the results of several calls to publish may be batched up in a single call to process. The process method receives a List<V> containing all intermediate results.

Let us put this mechanism to work for reading in a text file. As it turns out, a JTextArea is quite slow. Appending lines from a long text file (such as all lines in *The Count of Monte Cristo*) takes considerable time.

To show the user that progress is being made, we want to display the number of lines read in a status line. Thus, the progress data consist of the current line number and the current line of text. We package these into a trivial inner class:

```
private class ProgressData
{
   public int number;
   public String line;
}
```

The final result is the text that has been read into a StringBuilder. Thus, we need a SwingWorker<StringBuilder, ProgressData>.

In the doInBackground method, we read a file, a line at a time. After each line, we call publish to publish the line number and the text of the current line.

```
@Override public StringBuilder doInBackground() throws IOException, InterruptedException
{
   int lineNumber = 0;
```

```
    var in = new Scanner(new FileInputStream(file), StandardCharsets.UTF_8);
    while (in.hasNextLine())
    {
        String line = in.nextLine();
        lineNumber++;
        text.append(line).append("\n");
        var data = new ProgressData();
        data.number = lineNumber;
        data.line = line;
        publish(data);
        Thread.sleep(1); // to test cancellation; no need to do this in your programs
    }
    return text;
}
```

We also sleep for a millisecond after every line so that you can test cancellation without getting stressed out, but you wouldn't want to slow down your own programs by sleeping. If you comment out this line, you will find that *The Count of Monte Cristo* loads quite quickly, with only a few batched user interface updates.

In the process method, we ignore all line numbers but the last one, and we concatenate all lines for a single update of the text area.

```
@Override public void process(List<ProgressData> data)
{
    if (isCancelled()) return;
    var b = new StringBuilder();
    statusLine.setText("" + data.get(data.size() - 1).number);
    for (ProgressData d : data) b.append(d.line).append("\n");
    textArea.append(b.toString());
}
```

In the done method, the text area is updated with the complete text, and the Cancel menu item is disabled.

Note how the worker is started in the event listener for the Open menu item.

This simple technique allows you to execute time-consuming tasks while keeping the user interface responsive.

Listing 12.11 swingWorker/SwingWorkerTest.java

```
1 package swingWorker;
2
3 import java.awt.*;
4 import java.io.*;
```

(Continues)

Listing 12.11 *(Continued)*

```java
5  import java.nio.charset.*;
6  import java.util.*;
7  import java.util.List;
8  import java.util.concurrent.*;
9
10 import javax.swing.*;
11
12 /**
13  * This program demonstrates a worker thread that runs a potentially time-consuming task.
14  * @version 1.12 2018-03-17
15  * @author Cay Horstmann
16  */
17 public class SwingWorkerTest
18 {
19    public static void main(String[] args) throws Exception
20    {
21       EventQueue.invokeLater(() -> {
22          var frame = new SwingWorkerFrame();
23          frame.setDefaultCloseOperation(JFrame.EXIT_ON_CLOSE);
24          frame.setVisible(true);
25       });
26    }
27 }
28
29 /**
30  * This frame has a text area to show the contents of a text file, a menu to open a file
31  * and cancel the opening process, and a status line to show the file loading progress.
32  */
33 class SwingWorkerFrame extends JFrame
34 {
35    private JFileChooser chooser;
36    private JTextArea textArea;
37    private JLabel statusLine;
38    private JMenuItem openItem;
39    private JMenuItem cancelItem;
40    private SwingWorker<StringBuilder, ProgressData> textReader;
41    public static final int TEXT_ROWS = 20;
42    public static final int TEXT_COLUMNS = 60;
43
44    public SwingWorkerFrame()
45    {
46       chooser = new JFileChooser();
47       chooser.setCurrentDirectory(new File("."));
48
49       textArea = new JTextArea(TEXT_ROWS, TEXT_COLUMNS);
50       add(new JScrollPane(textArea));
51
```

```
52        statusLine = new JLabel(" ");
53        add(statusLine, BorderLayout.SOUTH);
54
55        var menuBar = new JMenuBar();
56        setJMenuBar(menuBar);
57
58        var menu = new JMenu("File");
59        menuBar.add(menu);
60
61        openItem = new JMenuItem("Open");
62        menu.add(openItem);
63        openItem.addActionListener(event -> {
64           // show file chooser dialog
65           int result = chooser.showOpenDialog(null);
66
67           // if file selected, set it as icon of the label
68           if (result == JFileChooser.APPROVE_OPTION)
69           {
70              textArea.setText("");
71              openItem.setEnabled(false);
72              textReader = new TextReader(chooser.getSelectedFile());
73              textReader.execute();
74              cancelItem.setEnabled(true);
75           }
76        });
77
78        cancelItem = new JMenuItem("Cancel");
79        menu.add(cancelItem);
80        cancelItem.setEnabled(false);
81        cancelItem.addActionListener(event -> textReader.cancel(true));
82        pack();
83     }
84
85     private class ProgressData
86     {
87        public int number;
88        public String line;
89     }
90
91     private class TextReader extends SwingWorker<StringBuilder, ProgressData>
92     {
93        private File file;
94        private StringBuilder text = new StringBuilder();
95
96        public TextReader(File file)
97        {
98           this.file = file;
99        }
100
```

(Continues)

Listing 12.11 *(Continued)*

```
101     // the following method executes in the worker thread; it doesn't touch Swing components
102
103     public StringBuilder doInBackground() throws IOException, InterruptedException
104     {
105        int lineNumber = 0;
106        try (var in = new Scanner(new FileInputStream(file), StandardCharsets.UTF_8))
107        {
108           while (in.hasNextLine())
109           {
110              String line = in.nextLine();
111              lineNumber++;
112              text.append(line).append("\n");
113              var data = new ProgressData();
114              data.number = lineNumber;
115              data.line = line;
116              publish(data);
117              Thread.sleep(1); // to test cancellation; no need to do this in your programs
118           }
119        }
120        return text;
121     }
122
123     // the following methods execute in the event dispatch thread
124
125     public void process(List<ProgressData> data)
126     {
127        if (isCancelled()) return;
128        var builder = new StringBuilder();
129        statusLine.setText("" + data.get(data.size() - 1).number);
130        for (ProgressData d : data) builder.append(d.line).append("\n");
131        textArea.append(builder.toString());
132     }
133
134     public void done()
135     {
136        try
137        {
138           StringBuilder result = get();
139           textArea.setText(result.toString());
140           statusLine.setText("Done");
141        }
142        catch (InterruptedException ex)
143        {
144        }
```

```
145         catch (CancellationException ex)
146         {
147            textArea.setText("");
148            statusLine.setText("Cancelled");
149         }
150         catch (ExecutionException ex)
151         {
152            statusLine.setText("" + ex.getCause());
153         }
154
155         cancelItem.setEnabled(false);
156         openItem.setEnabled(true);
157      }
158   };
159 }
```

---

javax.swing.SwingWorker<T, V> 6

- abstract T doInBackground()

  is the method to override to carry out the background task and to return the result of the work.

- void process(List<V> data)

  is the method to override to process intermediate progress data in the event dispatch thread.

- void publish(V... data)

  forwards intermediate progress data to the event dispatch thread. Call this method from doInBackground.

- void execute()

  schedules this worker for execution on a worker thread.

- SwingWorker.StateValue getState()

  gets the state of this worker, one of PENDING, STARTED, or DONE.

12.8 Processes

Up to now, you have seen how to execute Java code in separate threads within the same program. Sometimes, you need to execute another program. For this, use the ProcessBuilder and Process classes. The Process class executes a command in a separate operating system process and lets you interact with its standard input, output, and error streams. The ProcessBuilder class lets you configure a Process object.

NOTE: The `ProcessBuilder` class is a more flexible replacement for the `Runtime.exec` calls.

12.8.1 Building a Process

Start by specifying the command that you want to execute. You can supply a `List<String>` or simply the strings that make up the command.

```
var builder = new ProcessBuilder("gcc", "myapp.c");
```

 CAUTION: The first string must be an executable command, not a shell builtin. For example, to run the `dir` command in Windows, you need to build a process with strings `"cmd.exe"`, `"/C"`, and `"dir"`.

Each process has a *working directory*, which is used to resolve relative directory names. By default, a process has the same working directory as the virtual machine, which is typically the directory from which you launched the `java` program. You can change it with the `directory` method:

```
builder = builder.directory(path.toFile());
```

 NOTE: Each of the methods for configuring a `ProcessBuilder` returns itself, so that you can chain commands. Ultimately, you will call

```
Process p = new ProcessBuilder(command).directory(file)....start();
```

Next, you will want to specify what should happen to the standard input, output, and error streams of the process. By default, each of them is a pipe that you can access with

```
OutputStream processIn = p.getOutputStream();
InputStream processOut = p.getInputStream();
InputStream processErr = p.getErrorStream();
```

Note that the input stream of the process is an output stream in the JVM! You write to that stream, and whatever you write becomes the input of the process. Conversely, you read what the process writes to the output and error streams. For you, they are input streams.

You can specify that the input, output, and error streams of the new process should be the same as the JVM. If the user runs the JVM in a console, any user input is forwarded to the process, and the process output shows up in the console. Call

```
builder.redirectIO()
```

to make this setting for all three streams. If you only want to inherit some of the streams, pass the value

```
ProcessBuilder.Redirect.INHERIT
```

to the `redirectInput`, `redirectOutput`, or `redirectError` methods. For example,

```
builder.redirectOutput(ProcessBuilder.Redirect.INHERIT);
```

You can redirect the process streams to files by supplying `File` objects:

```
builder.redirectInput(inputFile)
   .redirectOutput(outputFile)
   .redirectError(errorFile)
```

The files for output and error are created or truncated when the process starts. To append to existing files, use

```
builder.redirectOutput(ProcessBuilder.Redirect.appendTo(outputFile));
```

It is often useful to merge the output and error streams, so you see the outputs and error messages in the sequence in which the process generates them. Call

```
builder.redirectErrorStream(true)
```

to activate the merging. If you do that, you can no longer call `redirectError` on the `ProcessBuilder` or `getErrorStream` on the `Process`.

You may also want to modify the environment variables of the process. Here, the builder chain syntax breaks down. You need to get the builder's environment (which is initialized by the environment variables of the process running the JVM), then put or remove entries.

```
Map<String, String> env = builder.environment();
env.put("LANG", "fr_FR");
env.remove("JAVA_HOME");
Process p = builder.start();
```

If you want to pipe the output of one process into the input of another (as with the | operator in a shell), Java 9 offers a `startPipeline` method. Pass a list of process builders and read the result from the last process. Here is an example, enumerating the unique extensions in a directory tree:

```
List<Process> processes = ProcessBuilder.startPipeline(List.of(
   new ProcessBuilder("find", "/opt/jdk-9"),
   new ProcessBuilder("grep", "-o", "\\.[^./]*$"),
   new ProcessBuilder("sort"),
   new ProcessBuilder("uniq")
));
```

```
Process last = processes.get(processes.size() - 1);
var result = new String(last.getInputStream().readAllBytes());
```

Of course, this particular task would be more efficiently solved by making the directory walk in Java instead of running four processes. Chapter 2 of Volume II will show you how to do that.

12.8.2 Running a Process

After you have configured the builder, invoke its start method to start the process. If you configured the input, output, and error streams as pipes, you can now write to the input stream and read the output and error streams. For example,

```
Process process = new ProcessBuilder("/bin/ls", "-l")
    .directory(Path.of("/tmp").toFile())
    .start();
try (var in = new Scanner(process.getInputStream())) {
    while (in.hasNextLine())
        System.out.println(in.nextLine());
}
```

 CAUTION: There is limited buffer space for the process streams. You should not flood the input, and you should read the output promptly. If there is a lot of input and output, you may need to produce and consume it in separate threads.

To wait for the process to finish, call

```
int result = process.waitFor();
```

or, if you don't want to wait indefinitely,

```
long delay = . . .;
if (process.waitfor(delay, TimeUnit.SECONDS)) {
    int result = process.exitValue();
    . . .
} else {
    process.destroyForcibly();
}
```

The first call to waitFor returns the exit value of the process (by convention, 0 for success or a nonzero error code). The second call returns true if the process didn't time out. Then you need to retrieve the exit value by calling the exitValue method.

Instead of waiting for the process to finish, you can just leave it running and occasionally call isAlive to see whether it is still alive. To kill the process, call destroy or destroyForcibly. The difference between these calls is platform-dependent. On UNIX, the former terminates the process with SIGTERM, the latter with SIGKILL. (The supportsNormalTermination method returns true if the destroy method can terminate the process normally.)

Finally, you can receive an asynchronous notification when the process has completed. The call process.onExit() yields a CompletableFuture<Process> that you can use to schedule any action.

```
process.onExit().thenAccept(
    p -> System.out.println("Exit value: " + p.exitValue()));
```

12.8.3 Process Handles

To get more information about a process that your program started, or any other process that is currently running on your machine, use the ProcessHandle interface. You can obtain a ProcessHandle in four ways:

1. Given a Process object p, p.toHandle() yields its ProcessHandle.
2. Given a long operating system process ID, ProcessHandle.of(id) yields the handle of that process.
3. Process.current() is the handle of the process that runs this Java virtual machine.
4. ProcessHandle.allProcesses() yields a Stream<ProcessHandle> of all operating system processes that are visible to the current process.

Given a process handle, you can get its process ID, its parent process, its children, and descendants:

```
long pid = handle.pid();
Optional<ProcessHandle> parent = handle.parent();
Stream<ProcessHandle> children = handle.children();
Stream<ProcessHandle> descendants = handle.descendants();
```

 NOTE: The Stream<ProcessHandle> instances that are returned by the allProcesses, children, and descendants methods are just snapshots in time. Any of the processes in the stream might be terminated by the time you get around to seeing them, and other processes may have started that are not in the stream.

The info method yields a ProcessHandle.Info object with methods for obtaining information about the process.

```
Optional<String[]> arguments()
Optional<String> command()
Optional<String> commandLine()
```

```
Optional<String> startInstant()
Optional<String> totalCpuDuration()
Optional<String> user()
```

All of these methods return `Optional` values since it is possible that a particular operating system may not be able to report the information.

For monitoring or forcing process termination, the `ProcessHandle` interface has the same `isAlive`, `supportsNormalTermination`, `destroy`, `destroyForcibly`, and `onExit` methods as the `Process` class. However, there is no equivalent to the `waitFor` method.

---

java.lang.ProcessBuilder 5

- `ProcessBuilder(String... command)`
- `ProcessBuilder(List<String> command)`

  constructs a process builder with the given command and arguments.

- `ProcessBuilder directory(File directory)`

  sets the working directory for the process.

- `ProcessBuilder inheritIO() 9`

  makes the process use the standard input, output, and error of the virtual machine.

- `ProcessBuilder redirectErrorStream(boolean redirectErrorStream)`

  if `redirectErrorStream` is true, the standard error of the process is merged into the standard output.

- `ProcessBuilder redirectInput(File file) 7`
- `ProcessBuilder redirectOutput(File file) 7`
- `ProcessBuilder redirectError(File file) 7`

  redirects the standard input, output, or error of the process to the given file.

- `ProcessBuilder redirectInput(ProcessBuilder.Redirect source) 7`
- `ProcessBuilder redirectOutput(ProcessBuilder.Redirect destination) 7`
- `ProcessBuilder redirectError(ProcessBuilder.Redirect destination) 7`

  redirects the standard input, output, or error of the process, where destination is one of:

  - `Redirect.PIPE`—the default behavior, access via the `Process` object
  - `Redirect.INHERIT`—the stream from the virtual machine
  - `Redirect.DISCARD`
  - `Redirect.from(file)`
  - `Redirect.to(file)`
  - `Redirect.appendTo(file)`

(Continues)

---

java.lang.ProcessBuilder 5 *(Continued)*

- Map<String,String> environment()

  yields a mutable map for setting environment variables for the process.

- Process start()

  starts the process and yields its Process object.

- static List<Process> startPipeline(List<ProcessBuilder> builders) 9

  starts a pipeline of processes, connecting the standard output of each process to the standard input of the next one.

---

java.lang.Process 1.0

- abstract OutputStream getOutputStream()

  gets a stream for writing to the input stream of the process.

- abstract InputStream getInputStream()
- abstract InputStream getErrorStream()

  gets an input stream for reading the output or error stream of the process.

- abstract int waitFor()

  waits for the process to finish and yields the exit value.

- boolean waitFor(long timeout, TimeUnit unit) 8

  waits for the process to finish, but no longer than the given timeout. Returns true if the process exited.

- abstract int exitValue()

  returns the exit value of the process. By convention, a non-zero exit value indicates an error.

- boolean isAlive() 8

  checks whether this process is still alive.

- abstract void destroy()
- Process destroyForcibly() 8

  terminates this process, either normally or forcefully.

- boolean supportsNormalTermination() 9

  checks whether this process can be terminated normally or must be destroyed forcefully.

- ProcessHandle toHandle() 9

  yields the ProcessHandle describing this process.

(Continues)

---

java.lang.Process 1.0 *(Continued)*

- CompletableFuture<Process> onExit() 9

  yields a CompletableFuture that is executed when this process exits.

---

java.lang.ProcessHandle 9

- static Optional<ProcessHandle> of(long pid)
- static Stream<ProcessHandle> allProcesses()
- static ProcessHandle current()

  yields the process handle(s) of the process with the given PID, of all processes, or the process of the virtual machine.

- Stream<ProcessHandle> children()
- Stream<ProcessHandle> descendants()

  yields the process handles of the children or descendants of this process.

- long pid()

  yields the PID of this process.

- ProcessHandle.Info info()

  yields detail information about this process.

---

java.lang.ProcessHandle.Info 9

- Optional<String[]> arguments()
- Optional<String> command()
- Optional<String> commandLine()
- Optional<Instant> startInstant()
- Optional<Instant> totalCpuDuration()
- Optional<String> user()

  yield the given detail information if available.

You have now reached the end of Volume I of *Core Java*. This volume covered the fundamentals of the Java programming language and the parts of the standard library that you need for most programming projects. We hope that you enjoyed your tour through the Java fundamentals and that you found useful information along the way. For advanced topics, such as the Java platform module system, networking, advanced user interface and graphics programming, security, and internationalization, please turn to Volume II.

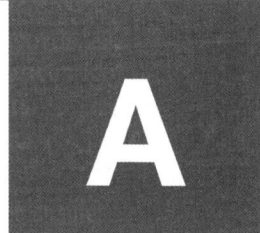

Appendix

This appendix lists all keywords of the Java language. Some keywords are "restricted." They have a special meaning only in certain circumstances (for example, in module declarations). Elsewhere, they can be identifiers.

Table A.1 Java Keywords

Keyword	Meaning	See Chapter
abstract	An abstract class or method	5
assert	Used to locate internal program error	7
boolean	The Boolean type	3
break	Breaks out of a switch or loop	3
byte	The 8-bit integer type	3
case	A case of a switch	3
catch	The clause of a try block catching an exception	7
char	The Unicode character type	3
class	Defines a class type	4
const	Not used	
continue	Continues at the end of a loop	3
default	The default clause of a switch, or a default method in an interface	3, 6

(Continues)

Table A.1 *(Continued)*

Keyword	Meaning	See Chapter
do	The top of a do/while loop	3
double	The double-precision floating-number type	3
else	The else clause of an if statement	3
enum	An enumerated type	3
exports	Exports a package of a module (restricted)	9 (Vol. II)
extends	Defines the parent class of a class, or an upper bound of a wildcard	4
final	A constant, or a class or method that cannot be overridden	5
finally	The part of a try block that is always executed	7
float	The single-precision floating-point type	3
for	A loop type	3
goto	Not used	
if	A conditional statement	3
implements	Defines the interface(s) that a class implements	6
import	Imports a package	4
instanceof	Tests if an object is an instance of a class	5
int	The 32-bit integer type	3
interface	An abstract type with methods that a class can implement	6
long	The 64-bit long integer type	3
native	A method implemented by the host system	12 (Vol. II)
new	Allocates a new object or array	3
null	A null reference (note that null is technically a literal, not a keyword)	3
module	Declares a module (restricted)	9 (Vol. II)
open	Modifies a module declaration (restricted)	9 (Vol. II)

(Continues)

Table A.1 *(Continued)*

Keyword	Meaning	See Chapter
opens	Opens a package of a module (restricted)	9 (Vol. II)
package	A package of classes	4
private	A feature that is accessible only by methods of this class	4
protected	A feature that is accessible only by methods of this class, its children, and other classes in the same package	5
provides	Indicates that a module uses a service (restricted)	9 (Vol. II)
public	A feature that is accessible by methods of all classes	4
return	Returns from a method	3
short	The 16-bit integer type	3
static	A feature that is unique to a class or interface, not to instances of a class	3, 6
strictfp	Use strict rules for floating-point computations	2
super	The superclass object or constructor, or a lower bound in a wildcard	5
switch	A selection statement	3
synchronized	A method or code block that is atomic to a thread	12
this	The implicit argument of a method, or a constructor of this class	4
throw	Throws an exception	7
to	A part of an exports or opens declaration (restricted)	9 (Vol. II)
throws	The exceptions that a method can throw	7
transient	Marks data that should not be persistent	2 (Vol. II)
transitive	Modifies a requires declaration (restricted)	9 (Vol. II)
try	A block of code that traps exceptions	7
uses	Indicates that a module uses a service (restricted)	9 (Vol. II)

(Continues)

Table A.1 *(Continued)*

Keyword	Meaning	See Chapter
var	Declares a variable whose type is inferred (restricted)	3
void	Denotes a method that returns no value	3
volatile	Ensures that a field is coherently accessed by multiple threads	12
with	Defines the service class in a provides statement (restricted)	9 (Vol. II)
while	A loop	3